THE UNOFFICIAL LEGO® MINDSTORMS® NXT
INVENTOR'S GUIDE

THE UNOFFICIAL LEGO® MINDSTORMS® NXT INVENTOR'S GUIDE

david j. **perdue**

no starch press

Printed in the United States of America

11 10 09   5 6 7 8 9

SUSTAINABLE FORESTRY INITIATIVE
Certified Fiber Sourcing
Label applies to the text stock   www.sfiprogram.org

ISBN-10: 1-59327-154-9
ISBN-13: 978-1-59327-154-1

Publisher: William Pollock
Production Editor: Megan Dunchak
Cover and Interior Design: Octopod Studios
Developmental Editor: William Pollock
Technical Editors: Tim Rueger and Christopher R. Smith
Copyeditor: Christina Samuell
Compositor: Riley Hoffman
Proofreader: Karol Jurado
Indexer: Nancy Guenther

The model shown on the cover, Tag-Bot, was built by Josh Pollock.

For information on book distributors or translations, please contact No Starch Press, Inc. directly:

No Starch Press, Inc.
555 De Haro Street, Suite 250, San Francisco, CA 94107
phone: 415.863.9900; fax: 415.863.9950; info@nostarch.com; www.nostarch.com

Library of Congress Cataloging-in-Publication Data

Perdue, David J.
  The unofficial LEGO Mindstorms NXT inventor's guide / David J. Perdue.
      p. cm.
  Includes index.
  ISBN-13: 978-1-59327-154-1
  ISBN-10: 1-59327-154-9
 1. Robots--Design and construction. 2. Robots--Programming. 3. LEGO toys. I. Title.
TJ211.P435 2007
629.8'92--dc22
                                    2007033731

about the author

David J. Perdue is a full-time student currently pursuing a degree in Management Information Systems from Liberty University. He is also the author of *Competitive MINDSTORMS: A Complete Guide to Robotic Sumo Using LEGO MINDSTORMS* (Apress, 2004), has written for *BrickJournal*, and was a featured speaker at the nationally recognized Texas Book Festival. In 2001, David bought his first MINDSTORMS® set, a Robotics Invention System, and has been given three Special Mention awards for his creations on the official online MINDSTORMS invention gallery. David also enjoys using LDraw computer-aided design software to create building instructions for his MINDSTORMS robots. David lives with his family in Round Rock, Texas, and maintains his website at http://www.davidjperdue.com.

technical reviewers

Tim Rueger is an avid LEGO® hobbyist and robotics enthusiast. He has worked as an analog chip designer at Cirrus Logic (formerly Crystal Semiconductor) in Austin, Texas, since 1997. He earned a Master of Science in Electrical Engineering from MIT in 1991. His wife, Shelley, graciously tolerates his varied choices in hobbies and his peculiar sense of humor. Tim enjoys mentoring home-school students in LEGO robotics competitions, including FIRST LEGO League, Robofest, and other local contests. His favorite LEGO creation is an all-LEGO floating-arm trebuchet capable of launching payloads over 40 feet. He hopes to someday defeat Steve Hassenplug in robotic sumo.

Christopher R. Smith (aka Littlehorn) is a senior quality assurance inspector in the Shuttle Avionics Integration Laboratory (SAIL) at NASA's Johnson Space Center in Houston, Texas. He has been designing LEGO MINDSTORMS robots since rediscovering the LEGO product in 1997, which led to joining the LEGO MINDSTORMS Developer Program for the NXT system. He enjoys contributing to the LEGO community by facilitating a supportive and challenging environment for enthusiasts worldwide.

I dedicate this book to my family: Jay, Mary Jo, Christopher, Stephanie, Tiffany, and Jessica.

brief contents

table of contents

PART I   INTRODUCTION TO LEGO MINDSTORMS NXT

PART II   BUILDING

PART III   PROGRAMMING

7
introduction to NXT–G... 67

8
advanced NXT–G programming... 91

PART IV   PROJECTS

acknowledgments

As I reflect on the entire writing process over the past year, I recognize more than ever the contributions that others have made to this book. First, I would like to offer a special thanks to William Pollock, founder of No Starch Press, for giving me the opportunity to write this book and for his advice on various aspects of it. I would also like to thank Josh Pollock, Bill's son, for assembling Tag-Bot (Chapter 14) so that No Starch's designer could photograph the robot for the front cover.

My technical reviewers, Tim Rueger and Christopher R. Smith, patiently read every chapter, built every robot, and tested every program. Thank you for your hard work and insightful comments. Thanks also go to my father, Jay Perdue, and Joe Meno for reviewing Chapter 4; Lutz Uhlmann for his excellent LGEO library and permission to use it for some of the computer-generated images in this publication; and the LEGO Company for its permission to use an image in Chapter 1.

I thoroughly enjoyed working with the staff at No Starch Press, and I'd like to thank them for their hard work in helping to develop this book. Elizabeth Campbell, Megan Dunchak, Christina Samuell, and Riley Hoffman especially played important roles in my project.

A big thanks goes to the LDraw community, whose excellent software tools I used to generate many of the images and all of the building instructions for this book. I would also like to specifically thank Michael Lachmann for his MLCad program; Kevin Clague for his LPub program; Travis Cobbs for his LDView program; and Tim Courtney, Steve Bliss, and Ahui Herrera, whose guide to LDraw, *Virtual LEGO* (No Starch Press, 2003), helped me to decide on the style for the building instructions in this book.

I'd also like to thank Melissa Follette, who took my picture for this book. I appreciate her time and effort as well as snapping some photos of one of the book's robots for me.

I'd like to *especially* thank my family for their patience, support, and words of encouragement that have been crucial to this book's completion. My parents, Jay and Mary Jo, my brother, Christopher, and my sisters—Stephanie, Tiffany, and Jessica—are not only my family but also my closest friends.

Finally, I recognize most of all that this book is the result of God working in my life. He faithfully guided me every step of the way and always strengthened me to take the next step. I offer this book to my Lord and Savior Jesus Christ. And I look forward to the day when I will meet face-to-face with the "author of life" (Acts 3:15).

introduction

When the LEGO Group commenced the MINDSTORMS series in 1998 with the release of the Robotics Invention System (RIS), it might have wondered what the outcome of LEGO-based robotics sets would be. Any ambiguity about the success of the MINDSTORMS series soon vanished, however, as the RIS sparked incredible sales, publicity, and enthusiasm around the world. Over the next eight years, MINDSTORMS inspired a multitude of fan websites, numerous robotic competitions, and more than 40 unofficial books based on the series.

During this time, the LEGO Group continued to expand the MINDSTORMS series with a variety of innovative products. The Robotics Discovery Set, Droid Developer Kit, Dark Side Developer Kit, Vision Command set, two updated versions of the RIS (1.5 and 2.0), and numerous expansion packs for the RIS made their appearances from 1998 to 2001. And then the flow of new MINDSTORMS products came to an abrupt halt.

As the years 2002 through 2005 failed to reveal significant developments within the MINDSTORMS product line, fans became concerned, eventually asking the unthinkable: Is MINDSTORMS being phased out? By the end of 2005, some fans had clearly given up hope.

Nevertheless, the LEGO Group had no intention of retiring the MINDSTORMS series. It had been hard at work developing yet another phenomenal product: the LEGO MINDSTORMS NXT set. LEGO finally revealed this secret project in January 2006 at the International Consumer Electronics Show, slating its release for fall of that year.

As the highly anticipated NXT set became available in various countries throughout the latter part of 2006, fans quickly recognized the set as the most powerful MINDSTORMS release to date. It implements many exciting features, including three motors with built-in rotation sensors (servo motors), four different sensors—touch, light, sound, and ultrasonic—the NXT microcomputer, hundreds of other building elements (i.e., LEGO pieces), the NXT-G programming software, and more.

about this book

The NXT set includes a user guide and instructions for several sample projects that are designed to help you learn how to create your own robots; however, this material only scratches the surface of the set's capabilities. By providing additional documentation and projects, *The Unofficial LEGO MINDSTORMS NXT Inventor's Guide* takes you beyond that material and helps you to more effectively and creatively use the NXT set.

Since different people have different levels of familiarity with MINDSTORMS, you might be wondering, "Is this book written for my experience level? Will it really help *me* to get the most out of my NXT set?" This book assumes no previous experience with LEGO or MINDSTORMS, but it also offers advanced material to challenge skilled MINDSTORMS users. In addition, this guide can serve as a helpful reference for all readers (for an example, see the NXT-G Quick Reference in Appendix B). The following overview of the book's contents, which are divided into four parts plus the appendixes, briefly shows you what you'll learn.

part I: introduction to LEGO MINDSTORMS NXT

Chapter 1   LEGO MINDSTORMS NXT: People, Pieces, and Potential

This chapter introduces you to the NXT set, discusses the three basic tasks involved in an NXT project—building, programming, and activating the robot—and describes the capabilities of the set.

Chapter 2   Getting Started with the NXT Set

This chapter guides you through the process of setting up your NXT set. You'll find instructions for installing the set's included software and exploring its main features; downloading programs to the NXT with the USB cable or Bluetooth technology; and properly organizing the LEGO pieces in the set.

part II: building

Chapter 3   Understanding the Electronic Pieces

The electronic pieces in the NXT set—the NXT microcomputer, servo motors, sensors, and electrical cables—are the most important ones. This chapter examines these vital pieces and their roles in a robot.

Chapter 4   Understanding the LEGO MINDSTORMS NXT Pieces

This foundational chapter examines the pieces in the NXT set as a whole, categorizing the different types and explaining their purposes. You'll also learn the name and color of each piece as well as how to measure them.

Chapter 5   Building Sturdy Structures

Since everyone wants to build robots that don't fall apart (unless you're intending to create a self-destructing robot), this chapter demonstrates essential techniques for building robust structures with beams from the NXT set.

Chapter 6   Building with Gears

This chapter provides a detailed discussion of how to effectively employ gears in your NXT creations. Using the variety of gears in the NXT set, you'll discover how to assemble gear trains and control gear train performance.

part III: programming

Chapter 7   Introduction to NXT-G

This chapter introduces you to NXT-G—the official programming language—and explores the NXT-G interface, some fundamental NXT-G concepts, and the Common palette that you'll use for basic programming.

Chapter 8   Advanced NXT-G Programming

Learning advanced programming skills can help you create more advanced robots, so this chapter covers advanced NXT-G features, including data wires, the Complete palette, and the Custom palette. You'll also learn how to expand the capabilities of NXT-G with Dynamic Block Update, an official add-on.

Chapter 9   Unofficial Programming Languages for the NXT

There are a number of unofficial languages that MINDSTORMS fans have developed for the NXT, and this chapter takes a brief look at four of them: NBC, NXC, leJOS NXJ, and RobotC.

part IV: projects

Chapter 10   The MINDSTORMS Method

This chapter presents the *MINDSTORMS method*, a simple but powerful strategy that guides you from the beginning of an NXT project to its completion. Understanding how to create your *own* robots will help you to better apply the concepts demonstrated by the robots in the following chapters.

Chapter 11   Zippy-Bot

Zippy-Bot is a two-wheeled vehicle with a ball caster. This easy-to-build robot that can zip around your room is the basis for the following two chapters.

Chapter 12   Bumper-Bot

Bumper-Bot explores your room and avoids objects by detecting them with its bumper, which is based on the touch sensor.

Chapter 13   Claw-Bot

Claw-Bot hunts for objects on the set's included NXT test pad and viciously pushes them away with the aid of a claw-like structure.

Chapter 14   Tag-Bot

Tag-Bot is a four-wheeled steering vehicle that plays flashlight tag. Shining a flashlight on Tag-Bot's light sensor for a short amount of time "tags" the robot, but Tag-Bot attempts to escape the light by dashing around the room.

Chapter 15   Guard-Bot

Guard-Bot is a fearsome, six-legged, walking and turning robot that can launch a ball at intruders. This creation can serve both as a stationary motion detector and as a guard that walks around the NXT test pad.

Chapter 16   Golf-Bot

Golf-Bot is a stationary robot that searches for a special target (which you'll also build with pieces from the NXT set), places a ball on the ground, and then hits the ball into the target. If the ball doesn't make it into the target, Golf-Bot will try again after you give it feedback through its sensors.

Appendix A   LEGO MINDSTORMS NXT Piece Library

This appendix presents images and detailed information about each type of piece in the NXT set.

Appendix B   NXT-G Quick Reference

You can flip to this appendix to find basic information about any of the standard NXT-G programming blocks.

Appendix C   Internet Resources

The final appendix lists useful Internet resources for learning more about LEGO MINDSTORMS NXT and LEGO in general.

companion website

I have created a companion website for *The Unofficial LEGO MINDSTORMS NXT Inventor's Guide* where you can download the source code, find updates, contact me, and more. The web address is http://www.nxtguide.davidjperdue.com.

PART I:

introduction to LEGO MINDSTORMS NXT

LEGO MINDSTORMS NXT:
people, pieces, and potential

Robots have the ability to grab people's attention. Whenever a robot makes an appearance—whether expected or unexpected—people quickly crowd around it, observing and commenting on its design, listening to the whir of its mechanical movement, and eagerly anticipating its next move.

Watching robots may be incredibly fun, but *creating* robots is even more exciting. For some time, the considerable expense and complexity of robotics prevented many people from pursuing it as a hobby, but a tidal wave of technological advancements has largely swept away these barriers. And riding this wave of technology is a particularly outstanding robotics kit: the LEGO MINDSTORMS NXT set (Figure 1-1).

The NXT set is a robotics toolset designed by the LEGO Group that empowers users to create functional robots entirely out of LEGO pieces. The set accommodates users of all ages, offering features that inspire and challenge children, teenagers, and adults alike. Generally speaking, an NXT project involves three tasks: *building* the robot, *programming* the robot, and then *activating* the robot. We'll briefly explore these topics in this chapter to gain a basic understanding of the NXT set.

Figure 1-1: The LEGO MINDSTORMS NXT set

building a robot

Including nearly 600 LEGO pieces, the NXT set offers plenty of building potential. But NXT robots aren't built with the famed bricks that have become synonymous with the word *LEGO*. (*Bricks* are a specific type of LEGO piece.) Throw back the lid of an NXT set (after admiring the impressive Alpha Rex robot on the front, of course), and you will reveal LEGO pieces that are almost anything but bricks.

Most prominent are the electronic elements: the NXT microcomputer, the motors, the sensors, and various lengths of electrical cables (Figure 1-2). In the center of Figure 1-2 is the *NXT microcomputer* (called simply *NXT*), which is powered by six AA batteries. Among its features are an LCD graphical display, four buttons for navigating the onscreen menu, three output ports, and four input ports. Using electrical cables, the motors connect to the NXT via the output ports and the sensors connect to the NXT via the input ports.

The NXT set includes three *servo motors*, motors with built-in rotation sensors, that make your robot come alive by enabling it to walk, drive, swivel, grasp, launch, and more. The NXT can use the built-in rotation sensors to precisely operate the motors, determining how many degrees their output shafts have turned to within one degree—an extremely useful feature.

Also included in the NXT set are four different sensors: a touch sensor, which provides your robot with a sense of touch by registering contact with other objects; a sound sensor, which provides your robot with a sense of volume; a light sensor, which provides your robot with a sense of light intensity; and an ultrasonic sensor, which provides your robot with a sense of sight by measuring distances between itself and other objects.

The NXT set also includes a wide variety of additional LEGO pieces, including gears, beams, axles, and pegs. Figure 1-3 shows just a few of these pieces, all of which work together with the electronic pieces to produce a complete, functional robot.

Figure 1-3: A selection of LEGO MINDSTORMS NXT pieces

Figure 1-2: The LEGO MINDSTORMS NXT electronic elements

While building robots with LEGO pieces requires no soldering, cutting, or customizing (you can simply use the pieces as they are), using those pieces *effectively* does require skill. To assist you in developing that skill, Part II covers MINDSTORMS NXT construction, teaching you about the individual pieces and showing you practical building techniques.

programming a robot

Although building may appear to be the primary feature of MINDSTORMS NXT, programming is a core feature as well. Programs are executed by the NXT and can tell your robot where to go, how to respond to its environment, how long to pause, which motor to power, which sensor to read, and so much more.

The NXT set provides a powerful yet easy-to-use programming environment called *NXT-G* (Figure 1-4) that is based on LabVIEW, the world-class graphical development tool made by National Instruments. NXT-G (which is compatible with both Windows and Mac OS X) takes a *drag-and-drop approach* to programming. This means that you drag and place programming blocks on your computer screen to create a program; you don't need to type any code. NXT-G offers a selection of basic programming blocks that helps beginners gradually develop their programming skills, as well as a number of advanced programming blocks that will be of use to more experienced users. Figure 1-4 shows a workspace that contains a single programming block, the *Move block*. By adjusting its options in the lower-left corner of the screen, you can use this block to power the motors in a variety of ways and make your robot move.

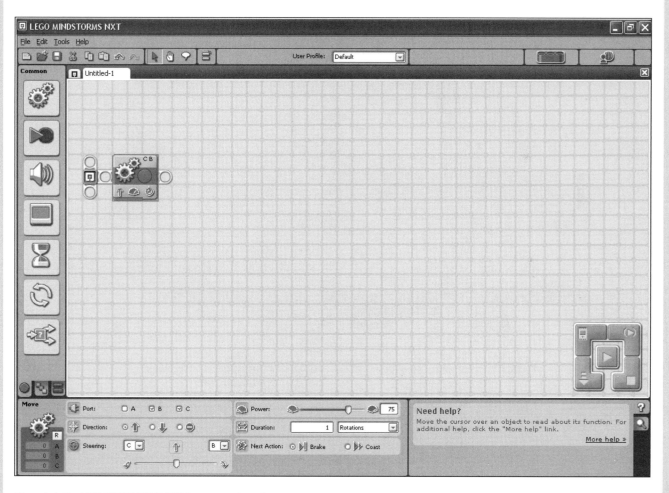

Figure 1-4: The LEGO MINDSTORMS NXT-G programming software

Figure 1-5 shows a complete NXT-G program for the Alpha Rex robot pictured on the front of the NXT set (the software bundled with the set includes step-by-step instructions for building the robot and designing this program). This program continually checks the value of a light sensor and initiates one of two responses depending upon the reading. If the reading is high (bright), the robot says, "Good morning!" and displays a smiley face on the NXT's LCD. If the reading is low (dim), the robot says, "Good night," begins to snore loudly, and displays Zs on the NXT's LCD. As this example demonstrates, a program doesn't have to involve physical movement to be interesting!

You download all NXT-G programs from your computer onto the NXT through either a USB cable (included with the NXT set) or a Bluetooth (wireless) connection. We'll discuss the details of this procedure in the next chapter.

Because programming is an essential (and interesting) aspect of MINDSTORMS, I've devoted Part III to it. There you'll find in-depth explanations of NXT-G and sample NXT-G programs. Following along with the examples and explanations will help you to quickly begin designing functional programs for your robots, even if you have never programmed before. In addition, we'll also touch upon several unofficial programming languages that LEGO fans

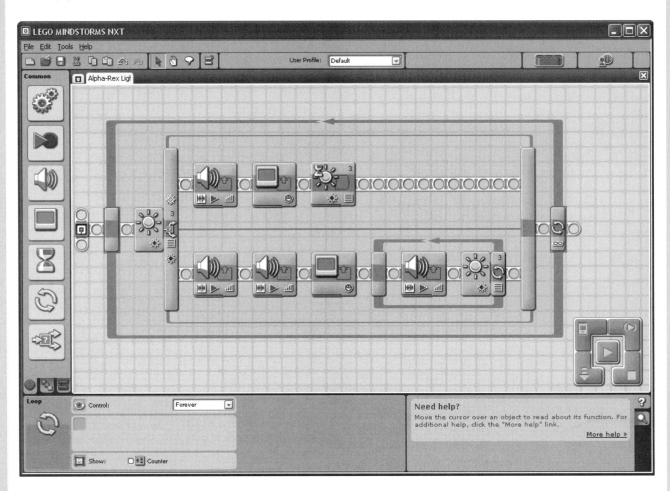

Figure 1-5: A complete NXT-G program for the Alpha Rex robot

have developed for the NXT. These alternative languages provide very different but powerful approaches to programming NXT robots, further expanding the capabilities of the NXT set.

activating a robot

Activating a robot is perhaps the most exciting part of a MINDSTORMS project. After investing time and effort in a robot's construction and programming, you finally get to watch your creation perform its task. If you need additional "equipment" to accommodate a robot's functionality (e.g., a robot that writes a message needs a pen and a sheet of paper), you can usually use common household items. Many times, however, activating an NXT robot can be as simple as placing it on the floor and pressing a button on the NXT. In any case, the only specialized testing element you'll need right away is already included in the NXT set: a test pad (Figure 1-6). Versatile and portable, this test pad is a handy robotics tool that we'll first use in Chapter 11.

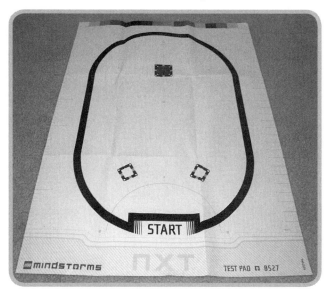

START

mindstorms   TEST PAD □ 8527

Figure 1-6: The MINDSTORMS NXT test pad

what can i create with MINDSTORMS NXT?

Building, programming, and activating robots are all fascinating and even instructive activities, but what kinds of robots can you actually create with the NXT set? Are you limited to the robot pictured on the front of the set, plus a few additional ones?

Not at all. The variety and quantity of pieces in the set allow you to create an almost unlimited number of robotic creations, whether creatures, vehicles, or crazy contraptions. The NXT set includes complete building instructions for several possible robots, including a walking scorpion that stings when disturbed, a robotic arm that can grab a ball and verbally announce its color, and the Alpha Rex humanoid, which is another walking (and dancing!) robot.

Part IV presents a variety of additional NXT projects—each one capable of being built just with the NXT set—and a method for creating NXT robots that will help you stay focused and organized when tackling your own projects. For the projects in this book, you'll also find step-by-step building instructions as well as accompanying NXT-G programs. As you build, program, and activate these robots in addition to those found in the NXT set, you will gain practical experience that will help you to successfully create your own robots.

conclusion

Having read this brief introduction to the LEGO MINDSTORMS NXT set, I imagine you're excited about creating NXT robots (if you weren't already). Of course, it's helpful to master a few basics before you set to work. After reading this chapter, you should have a firm grasp on exactly what the MINDSTORMS NXT set is, what sorts of tasks are involved in creating a robot with the NXT set, and what kinds of robots the NXT set can produce.

You're probably eager to begin using your NXT set, so in the next chapter I will guide you through the process of setting up and familiarizing yourself with your NXT set.

getting started with the NXT set

Opening a new LEGO MINDSTORMS NXT set is definitely exciting, but at the same time you're faced with the task of getting everything in order and then learning the basics. Sound difficult? It isn't. The NXT set's user-friendly features make getting started easy. And this chapter will further clarify that process, guiding you through the steps and offering additional insights and instructions that will help you to get the most out of your NXT set.

requirements

The NXT set requires six AA batteries and a computer (PC or Mac). In addition, if you choose to establish a wireless connection between your NXT and computer (an optional feature), your computer must have Bluetooth capability.

six AA batteries

Six AA batteries fit into the back of the NXT and power the NXT, motors, and sensors. Under most circumstances, regular alkaline batteries are recommended. If you will be using your NXT quite frequently, however, consider investing in some rechargeable batteries. They don't provide quite as much power as alkaline batteries do, but they can save you some money over the long haul.

a PC or mac

You'll use a computer—whether a desktop or a laptop—to access several of the set's features (such as NXT-G programming) after installing the included software. Windows and Mac OS X are compatible with the NXT software. Windows users must be running XP Professional, XP Home Edition with at least Service Pack 2 (SP2), or Windows Vista, while Mac users must be running OS X version 10.3.9, 10.4, or 10.5. A more detailed list of system requirements appears on the side of the NXT set and on the MINDSTORMS website (http://www.mindstorms.com).

bluetooth technology (optional)

Before you can download programs from your computer to your NXT, you must first establish a connection between the two. Although the NXT set includes a USB cable for making this connection, you can also use Bluetooth technology to establish a *wireless* connection. Your NXT is already Bluetooth capable, but your computer must either have built-in Bluetooth capability or use a Bluetooth adapter to wirelessly communicate with your NXT.

If your computer is not already Bluetooth capable, you should purchase a Bluetooth adapter from an electronics store or an online retailer such as Amazon.com. I *highly* recommend one particular adapter that is compatible with both Windows XP and Mac OS X: the D-Link DBT-120 (Figure 2-1). Don't let the image deceive you—this adapter is less than two inches long!

Figure 2-1: The D-Link DBT-120 Bluetooth adapter

NOTE The LEGO MINDSTORMS user guide packaged with the NXT set refers to a Bluetooth adapter as a *Bluetooth dongle*. These two terms—*Bluetooth adapter* and *Bluetooth dongle*—are basically synonyms.

While the DBT-120 works well with the NXT, it is important to note that not all Bluetooth adapters are compatible. If you cannot find this particular adapter, would like to purchase a different adapter, or want to determine if an adapter you already own is compatible, visit these websites for information regarding the performance of various Bluetooth adapters with the NXT:

* http://www.mindstorms.com/bluetooth
* http://www.vialist.com/users/jgarbers/
  nxtbluetoothcompatibilitylist

In the event that you will be purchasing an adapter other than the DBT-120, there is an easy (but not necessarily cheap) solution to finding a suitable adapter. As of this writing, the LEGO online store is now selling a Bluetooth adapter for $37.99. Browse to http://www.lego.com, search for *Bluetooth*, and select Bluetooth Dongle (Item #9847). Although the product description doesn't mention a specific brand, the adapter appears to be the Abe UB 22S, which is also compatible with both Windows XP and Mac OS X.

getting a quick start

What do you want to do immediately upon opening an NXT set? Naturally, you want to create a functional robot. With the quick-start kit included in your set (Figure 2-2), you can accomplish that within just 30 minutes. After opening your NXT set, grab this small kit and rip it open. The electronic elements, such as the NXT or motors, are packaged separately, but you will find in the kit everything you need to construct and test a small robot, including a little instruction booklet. Make sure you put six AA batteries in your NXT before you start building the robot (see the first page of the instruction booklet).

Figure 2-2: The MINDSTORMS NXT quick-start kit

NOTE Unlike the MINDSTORMS RIS set's RCX microcomputer, the NXT does *not* require you to initially download firmware onto it. Instead, the NXT ships with the firmware already installed, which it semi-permanently retains in its Flash memory. In other words, as soon as you put batteries in your NXT, it should be entirely operable.

After constructing and testing the robot according to the instruction booklet in the quick-start kit, you should then install the NXT software on your computer. This software contains the *challenges* (sample projects), the NXT-G programming environment, and more.

installing the NXT software

To begin installing the NXT software, place the CD-ROM included with your NXT set in your computer's CD-ROM drive. If you're running Windows XP, insert the CD-ROM and then wait for a menu like the one in Figure 2-3 to appear on the screen. Click the proper language (mine only offers English) and then follow the onscreen instructions.

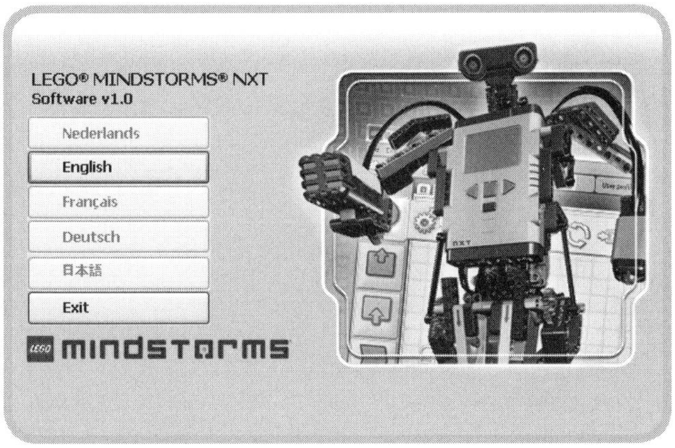

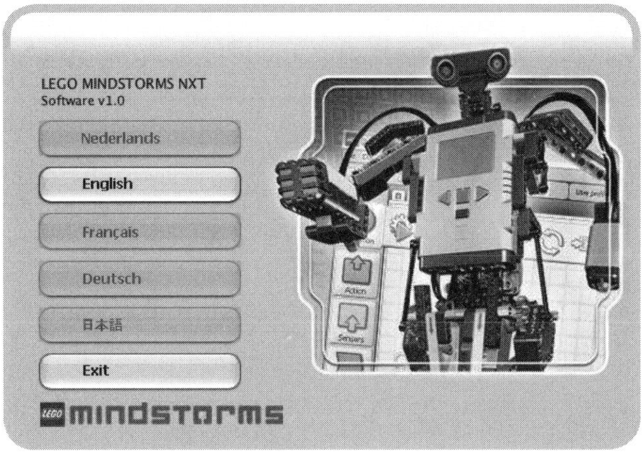

Figure 2-4: The installation menu for the NXT software (Mac OS X)

Figure 2-3: The installation menu for the NXT software (Windows XP)

NOTE Do *not* plug the USB cable that comes with the NXT set into your computer yet. You must plug it in *after* you have installed the software. You will learn more about this later in the chapter.

If you're running Mac OS X, you use a nearly identical menu to initiate the installation process (Figure 2-4). You must manually run the installer on the CD-ROM to view this menu, however. To do this, open the MINDSTORMS NXT CD-ROM on your computer and then double-click the **Install** icon. Once the menu appears, click the proper language and follow the instructions.

using the NXT software

Once you have installed the software, launch it and wait for the program to load. (It will take a little while to load the very first time, but it will load faster thereafter.) When it has finished loading, you will see the main screen (Figure 2-5). The NXT software has three main features: the NXT-G programming interface, the Robo Center, and My Portal.

NOTE I took all of the following screenshots in Windows XP. If you are running Max OS X instead, your screen will sometimes look slightly different, but you'll still be able to follow along.

getting started with NXT-G

Although Part III covers NXT-G in depth, in this chapter you'll learn how to tackle the following basic tasks so that you can begin programming:

* Create a new NXT-G program
* Add a programming block to an NXT-G program
* Establish a connection between your computer and the NXT using the USB cable or Bluetooth technology
* Download an NXT-G program to the NXT

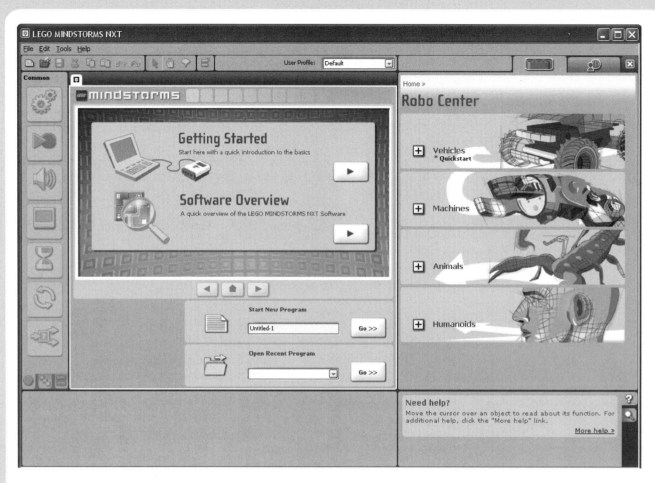

Figure 2-5: The main screen of the MINDSTORMS NXT software

Included in the NXT software is a short tutorial that covers these tasks (minus the Bluetooth technology), so we'll follow the tutorial's instructions. First, make sure you are still on the main screen of the NXT software. Click the large play button to the right of the words *Getting Started* (Figure 2-6) to begin watching the tutorial. Watch the entire tutorial before going forward.

creating a new NXT-G program

When the tutorial has finished, you can begin creating a new NXT-G program. Near the bottom of your screen you will see text boxes labeled with the words *Start New Program* and *Open Recent Program*. You can use these text boxes to create new NXT-G programs or quickly access existing ones. As instructed by the tutorial, you will create a new program called *First Program*. Type **First Program** into the Start New Program text box and then click the **Go>>** button (Figure 2-7).

Figure 2-6: Click the play button to begin watching the Getting Started tutorial.

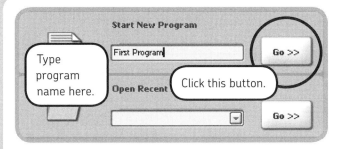

Figure 2-7: Creating the First Program

adding a programming block

Your new NXT-G program should appear right away (Figure 2-8). It's currently empty, containing no programming blocks, but you'll change that. Drag the **Sound block** into the

work area's Start position and adjust the block's sound level to **100** using the Volume slider in the block's configuration panel (Figure 2-9). When you've accomplished this easy task, your one-block program is complete.

In order to download the new program onto the NXT, you must establish a connection between your computer and the NXT. There are two ways to establish this connection: You can use the USB cable that came with your NXT set, or you can use Bluetooth technology to connect wirelessly. Since the tutorial uses the USB cable, let's begin with that method.

NOTE You can also consult page 9 of your LEGO MIND-STORMS user guide for information about using the USB cable.

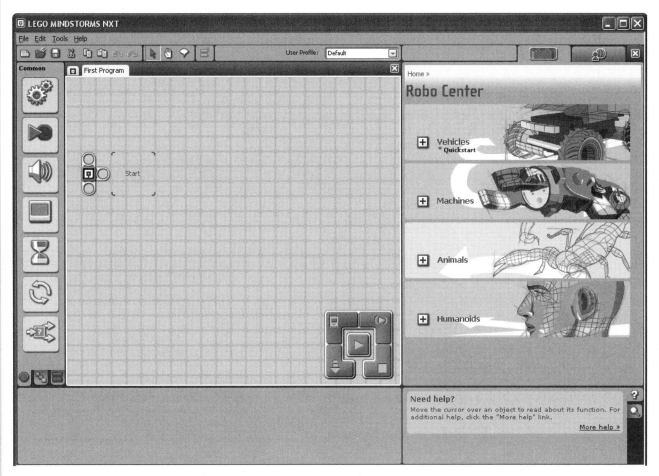

Figure 2-8: The blank First Program in NXT-G

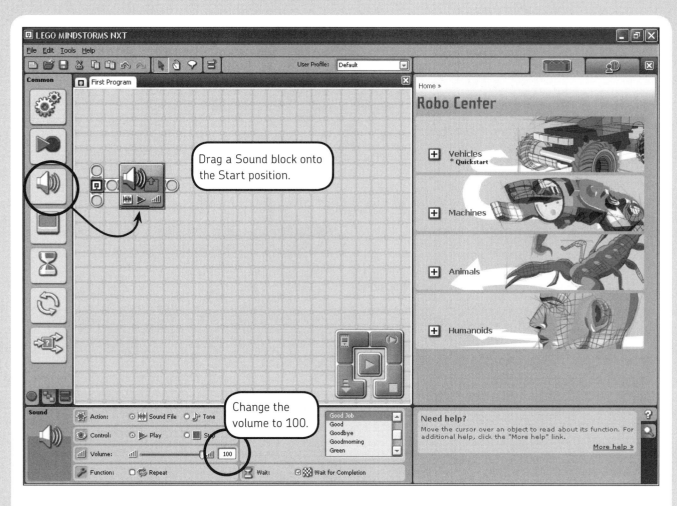

Figure 2-9: Adding and configuring a Sound block

establishing a connection with the USB cable and downloading a program

As instructed in the tutorial, press the orange button on your NXT to turn it on (you inserted the necessary batteries when you were working with the quick-start kit earlier). Grab the USB cable that came with your NXT set, plug the appropriate end into the USB port on the front end of your NXT, and then plug the other end into an available USB port on your computer. Only one end can fit into your NXT, so it will be easy to determine which end goes where. If your computer has more than one USB port, I encourage you to leave the USB cable plugged into it merely as a matter of convenience.

NOTE If you are using Windows XP, plugging in the USB cable will cause a small message about new hardware to appear in the lower-right corner of your screen. You shouldn't need to do anything except wait a moment until the message changes to *Your new hardware is installed and ready to use.*

Look to the bottom-right corner of your NXT-G program for the Controller. The Controller has five different buttons, but the only one we're interested in right now is the download button, the one in the lower-left corner (Figure 2-10) which downloads an NXT-G program to the NXT when clicked. Click it now to begin downloading your program. A dialog will appear to show the progress (Figure 2-11).

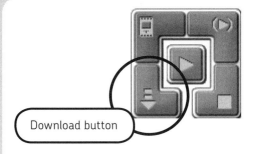

Download button

Figure 2-10: The download button on the NXT-G Controller

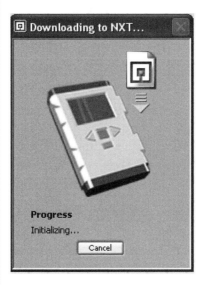

Figure 2-11: The program is downloading!

When the Downloading to NXT dialog disappears and your NXT beeps, the program has finished downloading. Remove the USB cable from your NXT, and then press the orange button on the NXT four times to select and run the First Program. The Sound block you used should result in your NXT saying, "Good job." Congratulations! You have just created, downloaded, and run your first NXT-G program.

If the download process for a program ever fails, check to make sure that the NXT is still powered on; it turns off by itself after a certain amount of time. If that's not the problem, check the NXT's LCD for the USB icon that denotes an unsuccessful USB connection (see page 10 of your LEGO MINDSTORMS user guide). If an unsuccessful USB connection is the culprit, try pulling out the USB cable and plugging it back in. If the USB connection seems fine, check the battery level indicator in the upper-right corner of the NXT's LCD to see how much power the batteries have—perhaps you need to put fresh batteries in the NXT.

If the battery level is fine, check the running icon adjacent to the battery level indicator on the NXT's LCD (see page 11 of your LEGO MINDSTORMS user guide). If the icon isn't spinning, your NXT has frozen and needs to be reset (see page 74 in your LEGO MINDSTORMS user guide for instructions).

If you are not going to set up Bluetooth technology right now, skip to "The Robo Center" on page 16.

establishing a connection with bluetooth technology and downloading a program

As mentioned earlier, your NXT can wirelessly communicate with your computer, but your computer must have either built-in Bluetooth capability or a Bluetooth adapter to use this feature. Although you must purchase the Bluetooth hardware for your computer separately, it is definitely worth the extra cost. Using Bluetooth technology, you can easily download a program to your robot from the other side of the room.

NOTE **Most computers do not have built-in Bluetooth capability. If your computer does or if you are already using a Bluetooth adapter, skip the next few paragraphs about installing Bluetooth adapters on your computer.**

At the beginning of this chapter, I recommended purchasing the D-Link DBT-120 Bluetooth adapter if your computer isn't already Bluetooth capable (see "Requirements" on page 9). It takes almost no effort to install this adapter on a computer running Windows XP or Mac OS X.

Plug the DBT-120 into one of your computer's available USB ports. Since Bluetooth software is included with Windows XP SP2 and Mac OS X 10.3.9 and 10.4, your computer should automatically install the adapter. You shouldn't have to do anything. If you are using Windows XP, message bubbles will pop up in the lower-right corner of your screen as the computer installs the adapter. The process will conclude with a familiar message: *Your new hardware is installed and ready to use.*

If for some reason your computer cannot automatically install the adapter, you should manually install it using the Bluetooth software provided with the DBT-120 on a CD-ROM (I have tested it and it is compatible). Just follow the instructions in the DBT-120 documentation to do so.

If you will be attempting to install a Bluetooth adapter other than the DBT-120, follow the same instructions. Plug the adapter into your computer and see if your computer will automatically install it. If it won't, install the Bluetooth software packaged with the adapter and see if that will work. It is *extremely important* that you plug the adapter into your computer *first* and see if your computer can automatically install it with its own Bluetooth software, however. The software packaged with some Bluetooth adapters does not work with the NXT, which means that those adapters require you to use your computer's built-in Bluetooth software.

Once your computer is Bluetooth capable, the next step is to connect your computer to your NXT. This process is accomplished within the NXT-G interface. The LEGO MINDSTORMS user guide contains excellent instructions for this procedure, so I will not duplicate them here. If you are using a PC, consult pages 29 through 31; if you are using a Mac, consult pages 31 through 33. All users should keep in mind these tips and reminders:

* By default, the Bluetooth feature of the NXT is turned off. As stated in the LEGO MINDSTORMS user guide, you can turn on this feature by going to the Bluetooth submenu on your NXT (see page 34 of the user guide) and turning on Bluetooth there (see page 35 of the user guide).
* Make sure that your NXT is visible to other Bluetooth devices. This option is also under the Bluetooth submenu on the NXT (see page 35 of the user guide).
* When you enter and confirm the passkey, remember that the default passkey is *1234* and that you will need to enter it both on your computer and on your NXT as illustrated in the LEGO MINDSTORMS user guide. (In case you're wondering, you will *not* need to enter a passkey every time you use the NXT's Bluetooth feature.)

Once you've gotten connected with Bluetooth technology, try downloading the same one-block program you used earlier. Downloading an NXT-G program with a Bluetooth connection is exactly the same as downloading one with a USB connection—only there is no cable involved! When the NXT uses its Bluetooth functionality, it does use more battery power, however. The following tips can help you conserve battery power:

* Turn off the NXT when it's not in use.
* Turn off the NXT's Bluetooth feature when you know you won't be using it in the near future.
* Configure the NXT's sleep mode to activate more quickly (see page 18 of the LEGO MINDSTORMS user guide).

the robo center

The *Robo Center* (Figure 2-12) is positioned on the far right side of the NXT software's main screen, and it contains eighteen challenges that are split up between four different robots: TriBot (a vehicle), RoboArm T-56 (a robotic arm), Spike (a walking and stinging scorpion), and Alpha Rex (a walking humanoid). Navigating within any of the four categories—Vehicles, Machines, Animals, and Humanoids—reveals challenge specifications, building instructions, and programming instructions.

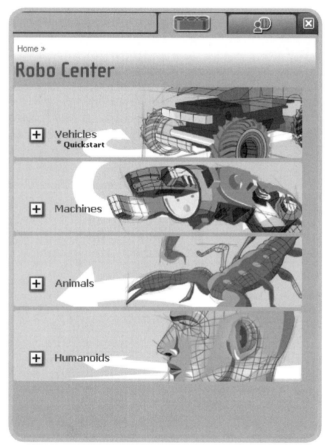

Figure 2-12: The Robo Center

NOTE In order to have more room to work in the NXT-G interface, you can temporarily hide the Robo Center by clicking the close button shown in the upper-right corner of Figure 2-12. You can reactivate the Robo Center by clicking the Robo Center tab, which appears at the top of the Robo Center and has a picture of an orange LEGO piece.

The Vehicles category includes the TriBot challenges (Figure 2-13). Do you remember the robot you created earlier with the quick-start kit? That was actually the TriBot's driving base. Since you have already constructed the driving base, you might consider working on one or more of the six TriBot challenges after reading this chapter to begin getting familiar with MINDSTORMS NXT construction and programming. Remember that you do not need to follow the building instructions for the driving base since you have already constructed it from the quick-start kit.

Figure 2-13: The TriBot challenges

my portal

My Portal is a feature that requires Internet access to display MINDSTORMS news and additional downloads and projects to complement the NXT set. When you launch it by clicking the My Portal tab adjacent to the Robo Center tab, the software connects to the MINDSTORMS website, gathers information, and then displays its dynamic content. Figure 2-14 shows what My Portal looks like as of this writing, but by the time you read this book, it will have almost certainly changed.

Figure 2-14: My Portal

organizing the MINDSTORMS NXT pieces

After working with the NXT software for a while, it's time to focus for a moment on the important task of organizing the MINDSTORMS NXT pieces. Open up your NXT set and look inside. All the pieces are jumbled together, which means that you will often have to search for a piece when you need it. And sometimes that search can be frustratingly long. To avoid needless rummaging, you need a much more efficient approach to working with the pieces.

In order to properly organize the pieces, you need some type of storage container with compartments or sections into which you can sort the pieces. If you have been a LEGO fan for some time, you may already have a storage system in place, allowing you to easily integrate the NXT pieces into your current collection of LEGO pieces. If you do not have such a setup, however, there are a number of solutions.

One of the best solutions is to purchase a single storage case or container capable of holding all of the NXT pieces. Figure 2-15 shows an example of such a case, which also has several noteworthy features: removable dividers for customizing the sizes of compartments, a sturdy handle, and a lid that is both transparent (it's nice to be able to see the contents at a glance) and lockable (you never know where your LEGO hobby might take you). You can find storage cases like this one at hardware and craft stores. In addition, Appendix C lists a website that offers storage cases specifically for the NXT set.

NOTE Due to the messy nature of the electrical cables, I always keep them individually bound with rubber bands when not in use. Simply take a single electrical cable, fold it compactly, and then bind it up with a rubber band.

On the other hand, it's not imperative that you fit all of the NXT pieces inside one storage container. In fact, if you are planning to expand your collection of LEGO pieces, it might be better to purchase several different storage containers that will accommodate more pieces. For many people, however, starting off with something simple and inexpensive may be the best approach; you can always expand your storage space later if necessary.

NOTE Before purchasing a storage case for the NXT set, you should first look around your home for one (or more) that you could use. Until I decided to remove them and replace them with LEGO pieces, the storage case in Figure 2-15 contained drill bits, nails, screws, and other hardware items.

As for how you should sort the NXT pieces within storage cases or containers, I offer no definitive answer. Providing one or more individual compartments for each type of piece is not always possible, and people use different sorting methods anyway. You will eventually develop your own personal approach to sorting LEGO pieces, too. Nevertheless, here are a few general tips for sorting the NXT pieces:

* Sort all of the electronic elements, including the cables, separately from the rest of the non-electronic pieces.

Figure 2-15: Plastic storage case containing all of the NXT pieces

* Keep miscellaneous pieces (such as the plastic balls) and pieces that you rarely use separate from the rest of the pieces.
* When working with relatively small quantities of pieces, you can sometimes successfully group similar pieces. You could group all or most of the gears together, for example.
* The NXT set contains several types of very small LEGO pieces that are quite similar, but you will find it helpful to keep most of these sorted separately.

conclusion

In this chapter, you set up your NXT set and became familiar with the most important of its features and functions. You began by using the quick-start kit to construct and test a small robot, and then you installed the NXT software on your computer and briefly explored the NXT-G interface, the Robo Center, and My Portal. Finally, you learned how to properly organize the NXT pieces for increased building efficiency. The next chapter begins Part II, which discusses the exciting topic of MINDSTORMS NXT construction.

PART II:

building

3

understanding the electronic pieces

Although the MINDSTORMS NXT set contains nearly 600 LEGO pieces, the handful of electronic pieces among these serve as the primary components, giving NXT creations their functionality and intelligence. What makes the electronic pieces even more significant is that their capabilities surpass those of all the other electronic LEGO elements that have appeared over the years. In order to effectively use these vital, powerful pieces in your creations, you must be familiar with their purposes and potential.

In this chapter, we'll examine the electronic NXT pieces, beginning with the most important one: the NXT. We'll go on to discuss the servo motors, the sensors, and finally the electrical cables that connect the electronic pieces.

the NXT

The LEGO Group has been developing microcomputers for some time. The RCX, Scout, Micro Scout, and other micro-computers have appeared in LEGO products since the late 1990s, but the NXT (Figure 3-1) is the most powerful LEGO microcomputer yet. Although the official documentation refers to it as an "intelligent, computer-controlled LEGO brick," it's hard to think of it as simply a special brick. While it's certainly a LEGO element, the NXT is also a genuine computer—a programmable microcomputer.

Dismantling the NXT would reveal the complexity of its design. A variety of electronics operate within it, but the primary components are a main processor and a co-processor that store information and manage the other electronics. The main processor is an Atmel 32-bit ARM7 pro-cessor with 256KB Flash memory, 64KB RAM, and a speed of 48 MHz. The co-processor is an Atmel 8-bit AVR processor with 4KB of Flash memory, 512 bytes of RAM, and a speed of 8 MHz. Although these processors cannot compare with those found in today's average home computer, remember that the NXT is a *micro*computer; for its purposes, it can operate with significantly less memory and processing speed.

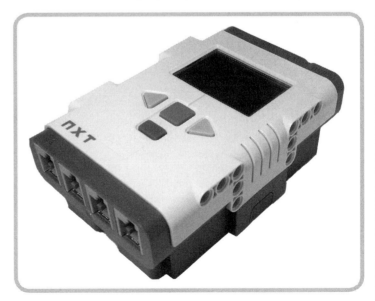

Figure 3-1: The NXT, a LEGO MINDSTORMS microcomputer

In the previous chapter, you installed batteries in your NXT, downloaded an NXT-G program to it, and ran the program. As you pressed buttons on the NXT to navigate to your NXT-G program, I'm sure you noticed the various menus on the NXT's LCD. You're probably also wondering about the features and capabilities of the NXT itself. In fact, the NXT has a number of noteworthy features:

* Three output (motor) ports
* A USB port
* Four input (sensor) ports
* Bluetooth capability
* A loudspeaker
* Four buttons and an LCD

output (motor) ports and the USB port

Although the output ports and the USB port serve entirely different purposes, we'll consider them together since they are located on the same end of the NXT (Figure 3-2).

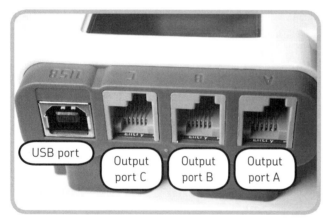

Figure 3-2: Output ports A, B, and C; and the USB port

Because the NXT is the only piece in the NXT set with its own power source (six AA batteries), it is generally responsible for powering and controlling the other electronic elements, including the motors. The *output ports* provide a place for the motors to connect to the NXT using electrical cables (which we'll discuss in "Connecting the Electronic Pieces with Electrical Cables" on page 28). All the necessary power and data are transmitted through the cables.

I've labeled the output ports and the USB port in Figure 3-2, but they are also labeled on the NXT. The output ports are labeled *A*, *B*, and *C*, and the USB port is labeled

USB. (The letters appear upside down in Figure 3-2 because they are meant to be viewed from the front of the NXT.) In the case of the output ports, the lettering scheme serves no purpose other than to help you distinguish between motors when using them. For example, suppose that you created a robotic arm and used a motor in output port A to raise and lower the arm and a motor in output port B to rotate the arm. When you're ready to program the NXT, you know to use output port A when changing the arm's position and output port B when rotating the arm.

NOTE You can use any of the motors in any of the output ports. The motors in the NXT set are identical, as are the output ports on the NXT itself.

The *USB port*, which you used in the previous chapter, allows you to use the USB cable included with the NXT set to establish a connection between your NXT and your computer. As Figure 3-2 shows, the USB port is square with a hump at the top. One of the ends of the USB cable matches this port, and the other end (which plugs into your computer) is shaped like a thin rectangle.

the input (sensor) ports

The NXT has four *input ports* labeled simply *1*, *2*, *3*, and *4* (Figure 3-3). Using electrical cables, sensors connect to the NXT through these ports. The cables transmit sensor data from the sensors to the NXT and any necessary power from the NXT to the sensors. Like the output ports' lettering scheme, the numbering scheme on the input ports serves no

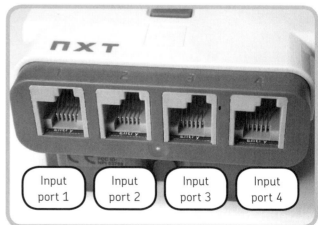

Figure 3-3: The input ports on the NXT

purpose other than to help you distinguish among the sensors when you're using them. You can use any input port for any sensor, but the default is as shown in Table 3-1.

table 3-1: the default sensor ports on the NXT

port number	sensor type
Port 1	Touch sensor
Port 2	Sound sensor
Port 3	Light sensor
Port 4	Ultrasonic sensor

NOTE The NXT is backward-compatible with the touch, light, rotation, and temperature sensors designed for the LEGO RCX microcomputer. In order to use these sensors with the NXT, you must purchase special converter cables, however. You can find these converter cables at the online LEGO store, where you can currently purchase them in packs of three for $9.99. Go to http://www.lego .com and search for *converter cable*.

bluetooth capability

One of the most exciting features of the NXT is its built-in Bluetooth capability. You learned in Chapter 2 that Bluetooth technology provides wireless connectivity (communication) to a wide variety of electronic devices. The NXT's Bluetooth functionality comes from a CSR BlueCore4 chip, which has its own microcontroller and an external 8MB Flash memory. It is entirely hidden from view, installed on the inside of the NXT, and has a range of approximately 10 m or 33 feet. Specifically, the NXT uses Bluetooth technology to accomplish the following:

* Wirelessly communicate with a Bluetooth-capable computer. Using the official NXT software on your computer, you can download programs to your NXT, manage files on your NXT, receive live feedback from your NXT, and more.
* Wirelessly communicate with other NXTs (up to three but one at a time). You could design separate creations that communicate with each other, or you could use multiple NXTs that communicate with each other in a single creation (a very *big* creation).
* Wirelessly communicate with a Bluetooth-enabled mobile phone. By downloading an application from the MINDSTORMS website and installing it on your mobile phone, you can use your phone to control an NXT robot! Visit http://mindstorms.lego.com/Overview/Bluetooth.aspx for more information.

a loudspeaker

Although not necessarily a vital component, the loudspeaker—which emits sound through several slits on the right side of the NXT—is both a useful and fun feature. When programming with NXT-G, you can select from an extensive list of sounds (including beeps, blips, music, words, phrases, numbers, and more) and play them over the NXT's loudspeaker. You can also use an interactive keyboard to specify tones for the loudspeaker to play. We'll explore this feature further when we discuss NXT-G in Part III.

NOTE LEGO fans have created unofficial software that can convert sound files on your computer to a format that is compatible with the NXT's loudspeaker. See Appendix C for a link to one of these resources.

In reality, the loudspeaker can either enhance a robot's performance or serve as the main feature of a robot's performance. Using the loudspeaker to play a round of applause when the robot has successfully accomplished a task would be an example of the former; using the loudspeaker to play tones based on sensor readings would be an example of the latter.

buttons and the LCD

Although sometimes overlooked, the NXT's four buttons and LCD are important features. Figure 3-4 shows the NXT's four buttons: the orange Enter button, the light gray Left and Right buttons, and the dark gray Clear/Go Back button. You use these buttons to navigate among and select the menus and options on the NXT's LCD. In addition, NXT-G allows you to program the Enter, Left, and Right buttons for use as sensors. An interesting application of this feature

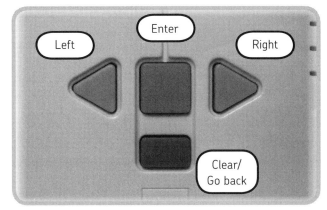

Figure 3-4: The four buttons on the NXT's interface

would be a robot that acts as a safe, requiring that you press the correct combination of buttons before allowing you to access its contents.

Figure 3-5 shows the NXT's LCD, which measures 100 × 64 pixels. Unlike other LEGO microcomputers, the NXT has an LCD capable of effectively displaying text and pictures. Figure 3-5 also points out some of the symbols and icons that appear above the line at the very top. For the details on all of the symbols and icons that display above the line, consult pages 10 through 11 of the LEGO MINDSTORMS user guide. For a flowchart showing the entire collection of menus, icons, and images that can appear on the menu (lower) portion of the LCD, see pages 20 through 21 of the LEGO MINDSTORMS user guide.

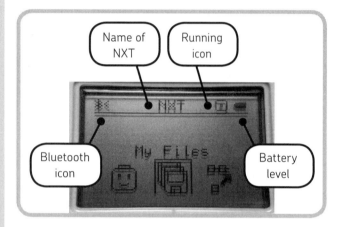

Figure 3-5: The NXT's LCD (main menu)

NOTE You can customize the name of your NXT. See page 11 of your LEGO MINDSTORMS user guide for instructions.

Don't think that you can only use the LCD to select programs or change options on the NXT. You can use it as an important part of your creations by displaying custom text and drawings on it. You can even use the LCD as the central feature of a robot. I've seen creations that use the LCD to play Tetris, Etch A Sketch, and tic-tac-toe!

Let's briefly practice navigating the NXT's menus and options. Turn on your NXT by pressing the orange Enter button. Your NXT will play a sound (via the loudspeaker), flash the LEGO and MINDSTORMS logos on the LCD, and then transition to the main menu.

Within the main menu are six submenus with different options and data (only three menus or selections are visible at a time). Use the Left and Right buttons to scroll through them. When you've found a menu that interests you, press the Enter button to select it. To return to the main menu (or the previous menu), press the Clear/Go Back button. The six submenus are as follows:

My Files This submenu contains all programs and sound files used by the NXT. The Software files subfolder holds the NXT-G programs downloaded from your computer, the NXT Files subfolder holds programs you've created on the NXT, and the Sound Files subfolder holds all sound files used by the NXT for itself and user-created programs.

NXT Program This submenu allows you to create programs directly on your NXT. Although definitely not as powerful as NXT-G, NXT Program is useful for making short programs for testing purposes. When selected, NXT Program presents five empty spaces on the LCD, and you fill the spaces with programming commands. See pages 14 and 15 of the LEGO MINDSTORMS user guide for more information.

View This submenu allows you to display the readings of the sensors and motors. For some of the electronic elements, you'll find more than one option (e.g., inches and centimeters for the ultrasonic sensor).[*] When you select an option, you'll be asked for the appropriate input or output port.

Bluetooth This submenu allows you to access and modify the Bluetooth options on your NXT. There are four subfolders—My Contacts, Connections, Visibility, On/Off—and a Search option that activates a search for other Bluetooth devices. Consult pages 34 through 35 of the LEGO MINDSTORMS user guide for more information on the Bluetooth submenu.

Settings This submenu allows you to view and change the settings of your NXT. Options include Volume, Sleep, NXT Version, and Delete Files.

Try Me This submenu offers five sample programs—one for the motor and one for each type of sensor—that you can use to test the electronic elements "in a fun way" (as the user guide puts it). When testing a motor or sensor with the Try Me programs, you must have the electronic element in the default output/input port on the NXT. The programs will remind you which port is the default.

---

[*] There are also options for viewing readings from sensors designed for the RCX microcomputer; these options are followed by an asterisk (*). As I mentioned earlier, you must use converter cables when using these sensors with the NXT.

the servo motors

NXT robots can move in many different ways—they can grasp, race, walk, swivel, and do much more—and that capability comes from using the NXT servo motors (Figure 3-6). Although LEGO motors have existed for some time, the three (identical) servo motors in the NXT set are entirely new.

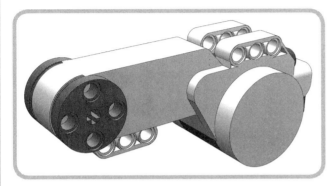

Figure 3-6: The MINDSTORMS NXT servo motor

The servo motor uses a built-in rotation sensor or *tachometer* to precisely control the motor's output shaft (leftmost part of Figure 3-6). Attached to the output shaft are two circular orange parts—which we'll call *shaft heads*—that rotate with the output shaft. By connecting LEGO pieces to the shaft heads, you can transfer power from the motor to your creation. The motor itself also has several places for attaching LEGO pieces, and you'll often use these to secure the motor to your creation.

When programming a motor, you can use degrees or rotations to specify a desired rotation. For example, you can specify that the motor should rotate 540 degrees (the motor is accurate to within 1 degree) or 1.5 rotations. Of course, you can also simply instruct the motor to run indefinitely or for a specified amount of time (e.g., 5 seconds). The motors can operate at different speeds, and there are certainly times when you'll want to run the motors at less than full power.

LEGO fans are heartily commending the introduction of the NXT servo motor, as previously they had to use a motor and a rotation sensor separately to achieve precision. Actually, many people have creatively used touch and even light sensors to count motor rotations!

the sensors

Much of the NXT set's potential lies in its excellent selection of sensors: a touch sensor, a light sensor, a sound sensor, and an ultrasonic sensor. These four sensors, which vary from extremely simple to highly complex, fall into two categories: passive sensors and digital sensors.

passive sensors

The NXT set includes three passive sensors: the touch sensor, the light sensor, and the sound sensor. *Passive sensors* either do not require power from the NXT in order to function or do not require the NXT to employ a process that rapidly switches between supplying power to and reading the value of a sensor.[*]

the touch sensor

Simple and robust, the touch sensor (Figure 3-7) is useful for a wide variety of applications. The orange missile-shaped tip at the front end, known as a *push button*, provides the touch-sensing abilities. When pressed, the push button moves backward into the sensor and completes an electrical circuit, resulting in a flow of electricity detectable by the NXT. When released, the push button springs forward, the circuit is broken, and the electrical flow stops. Hence, this simple configuration allows for only two possible conditions: pressed or released.

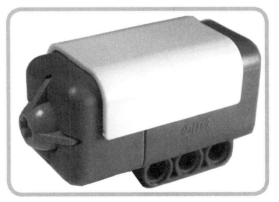

Figure 3-7: The MINDSTORMS NXT touch sensor

---

[*] The NXT also supports *active sensors*, which do require this rapid procedure. Examples of active sensors include the light and rotation sensors designed for the RCX microcomputer. You can only use these sensors with the NXT if you also use the converter cables mentioned earlier.

Just as the touch sensor can have two different conditions, it can produce two different values or readings. If the push button is pressed, the NXT reads the touch sensor as having a value of 1; if it's not pressed, the NXT reads a value of 0. You can use these values in several ways, however. For example, you can use any of the following conditions to trigger a reaction in a robot:

* When the touch sensor is pressed
* When the touch sensor is released
* When the touch sensor is pressed and released (i.e., "bumped")

On a historical note, the introduction of the NXT touch sensor ends the dominance of a previous touch sensor design. Though the NXT touch sensor is considerably less compact, I much prefer its more effective push button design. In addition, if you look closely at its push button, you'll notice a shaft-shaped hole. This hole accommodates LEGO axles, allowing you to customize the push button. (You'll learn about axles in Chapter 4.)

the light sensor

Capable of measuring light intensity, the NXT light sensor (Figure 3-8) can determine the brightness or darkness of its surrounding area as well as the light intensity of surfaces, enabling it to (indirectly) distinguish between surfaces of different colors. The NXT generates the light sensor's readings as a percentage. The highest possible reading is 100 percent, which you can easily achieve by holding the sensor up to a light bulb. The lowest possible reading is 0 percent, which you could achieve in a very dark closet.

Figure 3-8: The MINDSTORMS NXT light sensor

If you look at the front of the light sensor in Figure 3-8, you'll notice two small bulbs poking out. The one on top is a *phototransistor*, which measures the light; the one at the bottom is a *light-emitting diode (LED)*, which shines a bright red light. When the light sensor is positioned closely to a surface, the LED increases the amount of reflected light that the sensor reads, likewise increasing the sensor's sensitivity to different colors (different colors reflect light differently). You can turn off the LED in a program, however, if you want the light sensor to detect only the surrounding or ambient light.

Interestingly, the story behind the NXT light sensor is much the same as the one behind the NXT touch sensor. It was introduced along with the NXT set as a replacement for another long-standing version of light sensor, and it is also larger but more effective than the previous light sensor.

What are some practical ways to use the light sensor? Consider the official RoboArm T-56 robot, which picks up a ball, uses the light sensor to determine whether the ball is red or blue, and executes one of two actions, depending on the color of the ball. I personally enjoy creating line-following robots that use the light sensor to follow a line. Or perhaps you could create a robot that uses the light sensor to measure a room's overall light level and leaves the room when you turn off the lights. These are just a few possible applications for the light sensor.

the sound sensor

The NXT sound sensor (Figure 3-9), which is the first official MINDSTORMS sound sensor, can not only detect the volume of sound (*amplitude*) but also sound patterns. While it cannot distinguish between types of sound, such as between the sound of a bird and the sound of a cat, it still allows for many creative applications.

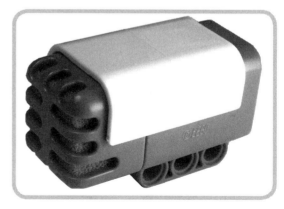

Figure 3-9: The MINDSTORMS NXT sound sensor

Although the NXT reads the sound sensor's value in decibels (*dB*, includes sounds inaudible to humans) or adjusted decibels (*dBA*, only sounds audible to humans), it reports the value as a percentage, with 100 percent being the highest and 0 percent being the lowest. According to page 23 of the LEGO MINDSTORMS user guide, a silent living room will return a value of 4 to 5 percent, distant talking will return a value of 5 to 10 percent, regular conversation or music playing at a moderate volume will return a value of 10 to 30 percent, and yelling or loud music will return a value of 30 to 100 percent.

NOTE 90 dB is the maximum sound level that the sound sensor can determine. The LEGO MINDSTORMS user guide says that this is "about the level of a lawnmower," but you must, of course, take into account the distance between the sound sensor and the object producing the noise.

The sound sensor definitely provides exciting possibilities for NXT robots. One of the simplest applications is a sound-activated robot that begins to operate and/or stops operating upon hearing a loud sound (such as a verbal command). Another possibility is an alarm robot that intently listens for intruders. Yet another example would be a robot that attempts to escape from sound by finding a more peaceful spot. Don't forget that the sound sensor can also recognize sound patterns, such as how many times you clap your hands.

digital sensor

The NXT set includes one digital sensor: the ultrasonic sensor. An NXT *digital sensor* has two important characteristics: It has its own microcontroller, which enables the sensor to take readings of its environment itself (as opposed to the NXT doing it), and it sends its data to the NXT using I^2C communication, which allows the sensor to operate independently and transmit only its readings to the NXT.[*]

NOTE Two other digital NXT sensors are now available: the compass sensor and the color sensor. The former enables a robot to determine its direction, and the latter enables a robot to identify a wide range of colors. You can currently purchase each sensor for $46.99 at the online LEGO store. Go to http://www.lego.com and search for *compass sensor* or *color sensor*.

Aside from the NXT itself, the ultrasonic sensor (Figure 3-10) is probably the most recognizable piece in the NXT set. This sensor detects an object by determining the distance between itself and that object, allowing your creations to "see," though certainly not to the extent that we humans can.

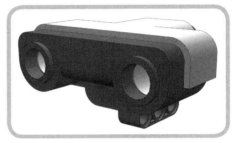

Figure 3-10: The MINDSTORMS NXT ultrasonic sensor

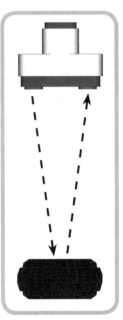

The ultrasonic sensor operates by employing a sonar-like system. It sends out a high-frequency sound wave (inaudible to humans), which reflects off an object. When the wave returns to the sensor (called a *rebound*), the sensor derives the distance between itself and the object from the amount of time it took the wave to return. Figure 3-11 demonstrates how the wave proceeds from the sensor, bounces off of an object, and returns to the sensor (the object is a LEGO tire from the NXT set).

Figure 3-11: The ultrasonic sensor sends and then receives a sound wave to determine the distance between itself and an object.

---

[*] See pages 7 and 9 of the LEGO MINDSTORMS Hardware Developer Kit (HDK) 1.00. For more information about this document, see "Getting NXTreme" on page 30.

The ultrasonic sensor has a range of 2.55 m or 8.4 feet and maintains an accuracy to within 3 cm. In fact, the NXT measures the ultrasonic sensor's readings in either inches or centimeters. An important fact to keep in mind is that the ultrasonic sensor can "see" larger objects better than smaller ones, objects with solid surfaces better than softer ones, and square or box-like objects better than smooth ones.

Multiple ultrasonic sensors operating in the same room can interfere with each other; this occurs when the waves sent by one ultrasonic sensor rebound to another ultrasonic sensor, resulting in an erroneous reading. This warning applies to ultrasonic sensors in general, not just the LEGO ones. You may not own more than one ultrasonic sensor at this point, but you should keep this warning in mind in case your robot will be performing against or around other NXT robots.

For many people, the ultrasonic sensor is the most exciting sensor in the NXT set. Possibly one of its most common applications is in object-avoiding robots. Unlike the setup in which a robot detects objects by bumping into them with a touch sensor, the ultrasonic sensor enables a robot to avoid objects *before* colliding with them. You can also use the ultrasonic sensor in many other creative ways. For example, you can use it as a motion detector, since an object that moves into or out of the sensor's range will result in different rebounds.

connecting the electronic pieces with electrical cables

I have repeatedly referred to the NXT electrical cables in this chapter, which you must use to connect the motors and sensors to the NXT. Now we shall finally examine these cables—one of which is shown in Figure 3-12—and then practice connecting the electronic elements.

Figure 3-12: A MINDSTORMS NXT electrical cable

First and foremost, there are seven cables in the NXT set, and you can use *any* cable to connect *any* electronic piece to the NXT: The cables are not specific to the electronic pieces. The only differences between the cables are their sizes. Table 3-2 lists the approximate lengths of the cables and the quantity and function of each size.

table 3-2: the electrical cables in the NXT set

length	quantity	function
20 cm/ 8 inches	1	This is the smallest size and perfect for connecting electronic pieces positioned closely to the NXT.
35 cm/ 14 inches	4	This is a moderate, all-purpose size that is the most commonly used.
50 cm/ 20 inches	2	This is the longest size, which you'll use in larger robots or situations requiring longer connections.

Second, observe Figure 3-13, a close-up of the connector on the ends of the NXT cables. If you're thinking that these connectors look very familiar—like the connectors on phone cable—you're correct. The connectors on the NXT cables are RJ12 connectors with a right-side adjustment. The prongs on the connectors can snap under excessive force, so you should treat the cables with a measure of care. They are, however, by no means fragile.

Figure 3-13: The transparent connector on NXT electrical cables

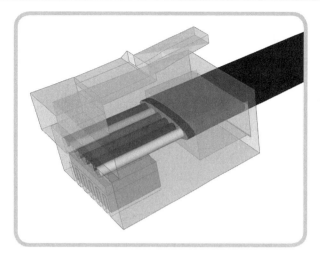

Figure 3-14: A computer-generated image of the connector more clearly reveals the six wires.

NOTE If you would like to purchase additional NXT cables, you can do so at the official online LEGO store. Go to http://www.lego.com, search for *connector cables*, and select Connector Cables for MINDSTORMS NXT (Item #8529). This package is currently selling for $9.99.

If you observe the center of Figure 3-13 carefully, you may be able to see the individual wires that run through the cables. There are actually six of these wires, and you can see them much more clearly in Figure 3-14, which is a computer-generated image. With these six wires, the electronic elements can transmit information to the NXT and receive power at the same time. For example, a servo motor can transmit its sensor data to the NXT (remember that it has a built-in rotation sensor) while also receiving power for rotating its output shaft.

Let's get some practice connecting and using the NXT electronics. Take out all of your electronic NXT pieces and press the Enter button to turn on the NXT. Take either end of an electrical cable and snap it into input port 1 on the NXT (it only fits one way). Take the touch sensor and snap the other end of the electrical cable into its connector hole in the back. Navigate to the Try Me menu on the NXT, select **Try-Touch**, and then select **Try-Touch Run** (you *must* have the touch sensor in input port 1 in order for this program to work). Press the touch sensor's push button, and the LEGO minifig pictured on the NXT's LCD will open his mouth and shout, "Whoops!" Release the push button, and the minifig's mouth will close.

When you are finished experimenting with the touch sensor, press the Clear/Go Back button on the NXT to exit the program. Now test the sound sensor, the light sensor, the ultrasonic sensor, and a servo motor, using the appropriate Try Me subfolders. Simply use an electrical cable to connect the sensor or motor to the NXT, navigate to the appropriate icon in the Try Me submenu, and run the program, making sure you have the sensor or motor on the default input or output port.

getting NXTreme

Where the NXT electronic pieces are concerned, we've only touched on the basics in this chapter. If you'd like to study them in depth, download the Hardware Developer Kit (HDK) from the MINDSTORMS website (http://mindstorms.lego .com/overview/nxtreme.aspx). The HDK contains detailed technical specifications and the schematics for all the electronic pieces. Although LEGO fans use it as a guide for developing custom hardware (e.g., custom sensors), you can also read it to simply gain a better understanding of how the NXT electronic elements operate. I used this excellent resource to help me write this chapter.

conclusion

In this chapter, you familiarized yourself with the NXT electronic pieces—the core components of the NXT set—and saw how they can transform static LEGO structures into active creations. We began by examining the powerful NXT, which is the most important electronic piece, and then we discussed the servo motors, which give your creations precise movement. We observed the four sensors in the NXT set—the touch, light, sound, and ultrasonic sensors—and then we concluded by discussing the NXT electrical cables used to connect the electronic pieces. In the next chapter, you will learn about the rest of the LEGO pieces in the NXT set—pieces that also serve important functions.

understanding the LEGO MINDSTORMS NXT pieces

Once you've begun creating your own robots with the NXT set, you'll soon ask a simple but significant question: "How do I build great NXT robots?" Obviously, the NXT set is capable of producing some impressive creations, but how do you utilize this potential? Is there a secret to constructing robust, functional, and remarkable robots? Not really. The key is simply to master the use of the LEGO pieces in the NXT set.

In the last chapter, we focused specifically on the electronic pieces. In this chapter, we'll broaden our scope to include all the pieces in the NXT set, addressing how to approach the entire system and then discussing the new pieces in detail. To acquire a real understanding of the pieces, we'll consider several basic questions: What types of pieces does the NXT set include and in what quantities? What are the names of these pieces? What are their purposes?

We'll build upon this knowledge in the next two chapters that discuss construction techniques.

introduction to the pieces

Since we'll be encountering dozens of different types of pieces in a variety of quantities, our first task should be to quickly get an overview of them. Figure 4-1 shows each type of piece included in the NXT set, followed by an *x* and a number, which specify the number of those pieces that are included.[*] Briefly look over it to gain a basic understanding of the building elements.

NOTE Pieces included in greater quantities are generally those that you'll use most often in your creations.

It's natural to assume that all of the pieces in the NXT set are MINDSTORMS pieces (i.e., pieces that are specific to MINDSTORMS), but besides the electronic pieces, most of them are actually LEGO *TECHNIC* pieces. Realizing this fact is important to understanding the nature of MINDSTORMS NXT construction. Launched in 1977, the TECHNIC series—previously known as the *Technical Sets* and then the *Expert Builder* series—enables you to create mechanical (but not intelligent) LEGO inventions. Because TECHNIC creations employ movement, they use many pieces that deviate from the standard brick-and-plate design. Over the years, TECHNIC has proven to be a particularly versatile and powerful subset of LEGO building.

---

[*] The exact count of pieces in your NXT set may slightly differ. A LEGO set usually includes a few extra of some of its smaller pieces.

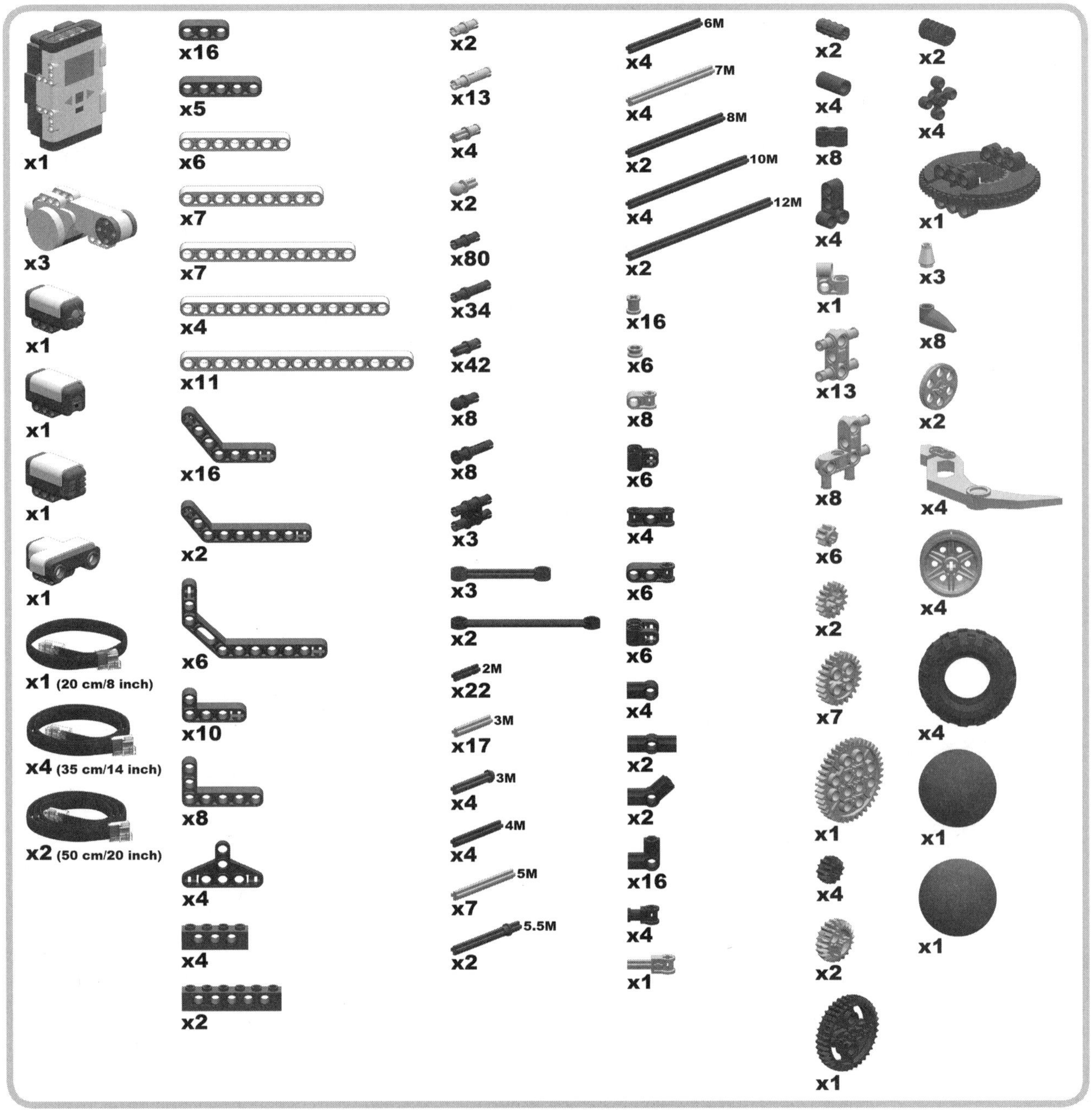

Figure 4-1: The types and quantities of pieces in the NXT set

In a sense, MINDSTORMS is an offshoot of TECHNIC because it relies heavily on TECHNIC pieces and building techniques. MINDSTORMS is actually more capable, however, because it combines the ingenuity of TECHNIC pieces with the power of its own robotic components. When using such a powerful construction system, it's particularly important that we begin by considering three related tasks: classifying the pieces, naming the pieces, and measuring the pieces.

classifying the pieces

First, we should *classify* the pieces—not only to stay organized, but also to develop a more complete understanding of
the pieces themselves. All of the pieces fit into five primary
categories; you'll soon learn which categories include which
pieces. The five main categories are as follows:

* Electronics
* Connectors
* Miscellaneous
  elements
* Beams
* Gears

naming the pieces

Second, we should *name* the individual pieces to facilitate
communication. Without names, trying to describe the
pieces would be a laborious (and sometimes humorous) task.
Imagine that I asked you to grab the *long, thin, shaft-like
piece that looks like a stick*. Using a term like *axle* instead is
much easier, isn't it?

The LEGO Group doesn't give each of its thousands of
pieces an official name, which is unfortunate but under-
standable. As a result, LEGO fans themselves have attached
names to the pieces, resulting in more than a little confusion
when the same piece goes by more than one name. Figure 4-2 illustrates this problem.

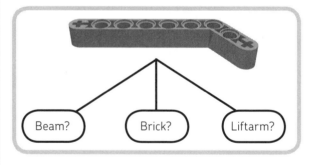

Figure 4-2: Should we call this piece a beam, a brick, or a liftarm?

I don't want to confuse you, so I have selected a unique
name for each piece in the NXT set and will *always* use these
names. I'll introduce them in this chapter and use them consistently throughout the book. You should realize, however,
that there isn't one naming standard that everyone follows,
and you'll almost certainly hear people refer to pieces by
names other than the ones I use in this book.

If you already have names with which you identify
TECHNIC pieces, feel free to continue using them. On
the other hand, you might consider adopting the naming
standard used in this book. I selected or created these

names after conducting considerable research, and I have
attempted to choose the most concise and accurate names.

measuring the pieces

Third, we should *measure* some of the pieces. You might be
thinking, "Why would I need to measure a LEGO piece? Isn't
a name all I need to identify a piece?"

That's a good question with a good answer. Because
many LEGO pieces are similar, it's sometimes necessary
to specify a piece's name *and* a measurement in order to
distinguish one piece from another. For example, imagine
that you're helping someone build a LEGO creation, and the
person extends part of the creation toward you and says,
"Make sure you use five straight beams on this section."

While this person has given you a specific name (you'll
learn about straight beams in a moment), you're also left
wondering, "What kind of straight beams? Small ones?
Medium ones? Large ones?" You wouldn't know and you
couldn't know. Figure 4-3 illustrates this problem.

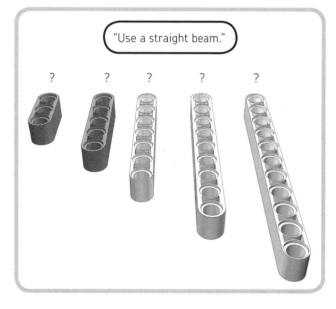

Figure 4-3: If you're told to use a straight beam, which kind of straight beam
should you choose?

Using some simple measurements resolves this issue
by allowing us to specify exactly which piece we're talking
about. For the most part, we'll use the module as our unit
of measure, but a few pieces in the NXT set use the LEGO
Unit. In addition, gears are often measured in their own way.
You'll learn the details of how and when to measure pieces
throughout the rest of this chapter.

NOTE A third criterion for identifying a piece is color; for example, you might refer to a *light stone gray straight beam*. Since we're only using the NXT set in this book, and each type of piece in the NXT set only comes in one color (with the exception of the plastic balls), piece colors generally don't present a problem.

examining the pieces

Armed with an understanding of the basic issues underlying the pieces in the NXT set, we're prepared to begin examining the five categories of pieces presented earlier: electronics, beams, connectors, gears, and miscellaneous elements. This is a fundamental section of the book that you should read thoroughly (and even reread), but don't feel like you have to digest it all at once. At any point, move on to something else if you would like—you can always come back to this section later.

NOTE Consult Appendix A for a summary of the attributes of each piece in the NXT set.

the electronics

This first category includes the NXT, the three servo motors, the four sensors, and the electrical cables. Because of these elements' complexity and capability, I devoted Chapter 3 to them and will not discuss them in further detail here.

the beams

The second category to consider is the beams category. The term *beam* encompasses a variety of pieces that compose the structures of creations. In other words, beams are to your LEGO creations what a foundation, walls, and a roof are to a house. Figure 4-4 offers a comprehensive view of the various types of beams in the NXT set; match up the numbers above each of the pieces with the numbers in Table 4-1 for information about each piece.

We can break down these beams into four subcategories:

* Straight beams   * Angled beams
* Half-beams   * TECHNIC bricks

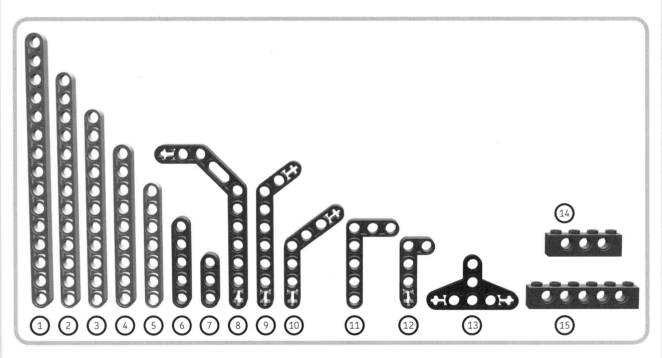

Figure 4-4: The beams in the NXT set

table 4-1: the NXT beams

number in figure 4-4	piece name	piece color (in NXT set)
1	15M (straight) beam	Light stone gray
2	13M (straight) beam	Light stone gray
3	11M (straight) beam	Light stone gray
4	9M (straight) beam	Light stone gray
5	7M (straight) beam	Light stone gray
6	5M (straight) beam	Dark stone gray
7	3M (straight) beam	Dark stone gray
8	11.5M angled beam	Dark stone gray
9	9M angled beam	Dark stone gray
10	7M angled beam	Dark stone gray
11	7M perpendicular angled beam	Dark stone gray
12	5M perpendicular angled beam	Dark stone gray
13	Triangular half-beam	Black
14	1 × 4 TECHNIC brick	Dark stone gray
15	1 × 6 TECHNIC brick	Dark stone gray

the straight beam

The *straight beam* (Figure 4-5) is the most basic structural piece, which means that you'll use it often. It has a smooth exterior, rounded ends, and an odd number of holes called *round-holes* that run along the middle. These round-holes are chiefly used to connect the beam to other pieces with TECHNIC connectors (which we'll discuss later in this chapter).

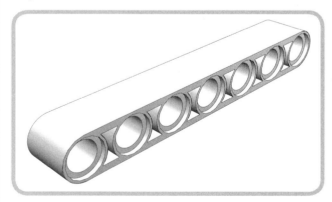

Figure 4-5: The 7M (straight) beam

If you observe Figure 4-4 again, you'll notice seven different types of straight beams in the NXT set. To distinguish one straight beam from another, we measure their lengths in modules, a basic TECHNIC unit that is abbreviated *M*. Between two adjacent round-holes on a straight beam is an hourglass-shaped depression. A *module* is the distance from the center of one of these depressions to the center of an adjacent depression, and it measures approximately 8 mm. Figure 4-6 shows exactly what a module is, and Figure 4-7 shows how to use the module to measure a straight beam. In "The Connectors" on page 37, we'll also use the module to measure other types of pieces.

NOTE The number of round-holes in a straight beam corresponds to its module measurement, which means you can count round-holes as a measuring shortcut. For example, a straight beam with five round-holes has a module measurement of 5M.

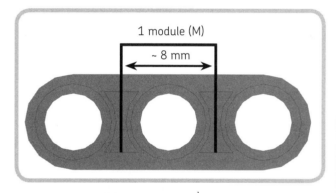

Figure 4-6: A module (M) is about 8 mm, the distance from the center of one hourglass-shaped depression to the center of the adjacent depression.

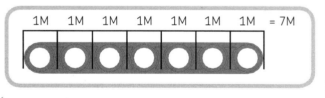

Figure 4-7: Add up the number of modules to get a total measurement of 7M.

To properly describe a straight beam, you must give both its module measurement and its name. However, when we give a straight beam's module measurement, we drop the word *straight* from the name. For example, a straight beam measuring 3M (three modules) would be called a *3M beam*, a straight beam measuring 5M (five modules) would be called a *5M beam*, and so on. When only the module measurement and the word *beam* are given, it's understood that the piece in question is a straight beam.

Straight beams exist in sizes ranging from 2M to 15M, but the seven types of straight beams in the NXT set range in sizes from 3M to 15M. Of course, the different sizes are designed to accommodate different situations: In one case, you may want to use a long straight beam; in another situation, you may want to use a short straight beam.

the angled beam

The *angled beam* (Figure 4-8) is primarily different from the straight beam in that one or more sections of the beam are angled. Sometimes this type of beam simply makes a creation more interesting, while other times it can play important structural roles (e.g., some angled beams work well as "fingers" on grabbing mechanisms). Looking back at Figure 4-4 once again, you'll notice that five types of beams in the NXT set fall into the angled beam subcategory, ranging in sizes from 5M to 11.5M.[*] Included among these are two types of *perpendicular angled beams*, which are beams angled at exactly 90 degrees.

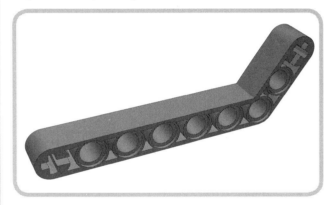

Figure 4-8: The 9M angled beam

In the NXT set, most of the angled beams have another important but less noticeable characteristic: cross-holes. Figure 4-9 shows the same beam as in Figure 4-8 but points out its two cross-holes. A *cross-hole* is specifically used with connectors known as *cross-axles* or simply *axles*, which you'll learn about in "The Connectors" on the next page. When measuring an angled beam with or without cross-holes, proceed exactly as you would when measuring a straight beam (Figure 4-10).

---

[*] The 11.5M angled beam has a half module in its measurement because of a 1.5M gap between two round-holes.

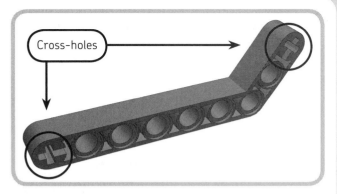

Figure 4-9: Some angled beams, such as this 9M angled beam, have cross-holes.

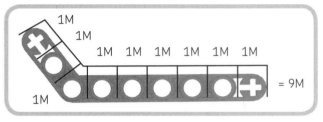

Figure 4-10: Measure an angled beam just as you would measure a straight beam.

the half-beam

A *half-beam* is simply a beam that is half the width (4 mm) of a regular beam (8 mm). A variety of these half-beams exist, but the NXT set contains only one kind: the triangular half-beam (Figure 4-11). Uniquely shaped and possessing both round-holes and cross-holes, the *triangular half-beam* enables you to solve structural and mechanical problems in unconventional ways. We don't measure the triangular half-beam.

Figure 4-11: The triangular half-beam

the TECHNIC brick

The last type of beam is the TECHNIC brick (Figure 4-12).* Because it has round-holes like the other beams, the *TECHNIC brick* can interact with other TECHNIC pieces by means of connectors. On the other hand, because it has studs, the TECHNIC brick can interact with other studded LEGO pieces, such as bricks. *Studs* are small cylindrical "bumps" on top of certain LEGO pieces that can snap into the bottom of other studded pieces—a connection system known as the *stud-and-tube coupling system*. There are only a handful of TECHNIC bricks in the NXT set, so they do not play a large role in NXT construction. Nevertheless, there are some situations in which you'll find them very useful.

Figure 4-12: The 1 × 4 TECHNIC brick has four studs.

We don't measure TECHNIC bricks with the module but rather with the LEGO Unit, which is also used to measure bricks and plates. In simplified terms, the *LEGO Unit* measures width and length by counting studs. How many studs wide is the TECHNIC brick in Figure 4-12? It's one stud wide. How many studs long is it? It's four studs long. Take the two measurements, combine them, put the width before the length, and you get what's called a 1 × 4 TECHNIC brick. (Note that 1 × 4 is pronounced *one by four*.) As this example demonstrates, using the LEGO Unit is very intuitive.

the connectors

We can now transition to the connectors category, which is the largest category in terms of both types and quantities of pieces. *Connector* is a general term which encompasses a variety of pieces that provide connectivity. In essence, TECHNIC connectors are like nails, staples, screws, bolts, and other similar items that hold a structure together. Figure 4-13 presents the various connectors in the NXT set; match up the numbers by the pieces with the numbers in Table 4-2 for information about each piece.

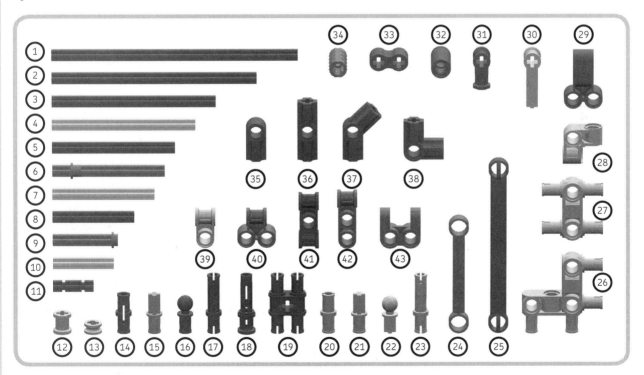

Figure 4-13: The connectors in the NXT set

---

* It's sometimes also known as the *studded beam*, but the name *TECHNIC brick* is more commonly used.

table 4-2: the NXT connectors

number in figure 4-13	piece name	piece color (in NXT set)
1	12M axle	Black
2	10M axle	Black
3	8M axle	Black
4	7M axle	Medium stone gray
5	6M axle	Black
6	5.5M stopped axle	Dark stone gray
7	5M axle	Medium stone gray
8	4M axle	Black
9	3M studded axle	Dark stone gray
10	3M axle	Medium stone gray
11	2M notched axle	Black
12	Bushing	Medium stone gray
13	Half-bushing	Medium stone gray
14	Friction peg	Black
15	Friction axle peg	Blue
16	Friction ball peg	Black
17	3M friction peg	Black
18	Bushed friction peg	Black
19	Double friction peg	Black
20	Peg (smooth)	Medium stone gray
21	Axle peg (smooth)	Tan
22	Axle ball peg (smooth)	Medium stone gray

number in figure 4-13	piece name	piece color (in NXT set)
23	3M peg (smooth)	Medium stone gray
24	Steering link	Dark stone gray
25	9M steering link	Black
26	5M pegged perpendicular block	Medium stone gray
27	3M pegged block	Medium stone gray
28	Cornered peg joiner	Medium stone gray
29	Double peg joiner	Black
30	Extended catch	Medium stone gray
31	Catch	Black
32	Peg extender	Black
33	Flexible axle joiner	Black
34	Axle extender	Dark stone gray
35	#1 angle connector	Black
36	#2 angle connector	Black
37	#4 angle connector	Black
38	#6 angle connector	Black
39	Cross block	Medium stone gray
40	Double cross block	Dark stone gray
41	Inverted cross block	Black
42	Extended cross block	Dark stone gray
43	Split cross block	Dark stone gray

We can break down these connectors into three subcategories:

* Axles
* Pegs
* Connector blocks

the axles

The *axle* is one of the most vital connectors, but it's nothing more than a cross-shaped shaft (Figure 4-14). Although its full name is the *cross-axle*, it's more commonly known simply as the *axle*, which is how I'll refer to it. The NXT set includes 72 axles of 11 different types, which signals that the axle is indeed an important piece.

Figure 4-14: The 8M axle

I mentioned earlier that the cross-holes in beams (and other pieces) specifically accommodate axles, so you might think that you only use axles in situations involving cross-holes. Using an axle with one or more cross-holes does create a very rigid connection, as the leftmost part of Figure 4-15 demonstrates. However, using an axle with one or more round-holes allows the axle to spin freely, as the rightmost part of Figure 4-15 demonstrates. (Note that we would normally keep the axle in place with other pieces.) Powered by motors, rotating axles are the basis of nearly all forms of movement in MINDSTORMS creations. By attaching one or more pieces to rotating axles, we can develop various forms of movement, such as driving or walking. You'll learn more about this concept in "The Gears" on page 42 and in Chapter 6.

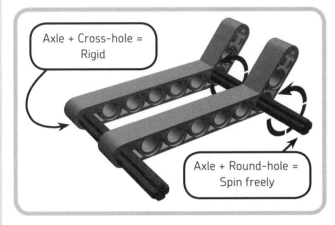

Figure 4-15: Using an axle in cross-holes creates a very rigid connection, while using an axle in round-holes allows the axle to spin freely.

Since a variety of axles exist—mainly in different lengths—it's imperative that we measure them. The module is the unit of measure for axles, but it's more difficult to measure axles in modules because axles don't have round-holes or cross-holes. Fortunately, in the NXT set and other more recent LEGO sets, the axles are color-coded: All axles with an even module measurement (2M, 4M, 6M, and so on) are black, while all axles with an odd module measurement (3M, 5M, 7M, and so on) are medium stone gray. This means that with some practice you can successfully deduce an axle's module measurement just by its color and relative length. If you're ever unsure of an axle's size, you can also compare it against the axles pictured on the back cover of your LEGO MINDSTORMS user guide.

NOTE Two axles are neither black nor medium stone gray. The 5.5M stopped axle and the 3M studded axle are dark stone gray. This difference in color merely signifies that these are specialized axles. Experimenting with them will reveal just how they're different from ordinary axles.

Finally, I must mention two important pieces that we call *axle accessories*: the bushing and half-bushing (Figure 4-16). These two parts, which are essentially cross-holes in piece form, rigidly hold their place anywhere along an axle. They generally function as separators when positioned between pieces on an axle and as fasteners when used to prevent an axle from falling out of a round-hole. You'll always want to keep some of these pieces close at hand when working with axles.

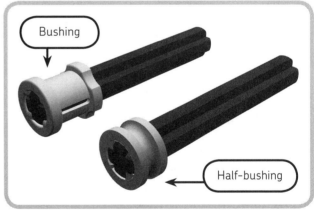

Figure 4-16: The bushing and half-bushing are assistants to the axle.

the pegs

Though quite small, pegs are also vital components of TECHNIC construction (Figure 4-17).[*] They can be used to easily yet firmly connect two or more pieces. The NXT set includes nearly 200 pegs of 10 different types—that's about 35 percent of its entire collection of pieces! Depending on the type of peg, it may snap into a round-hole, a cross-hole, or both. The peg shown in Figure 4-17 is the most basic peg—in fact, it's called *the peg*—and when pushed into a round-hole, it goes as far as its *stop ridge*, which circles the middle of the peg. Hence, it can connect two pieces, one on each side of its stop ridge. However, we often use two or more pegs together, as Figure 4-18 illustrates.

Figure 4-17: The peg

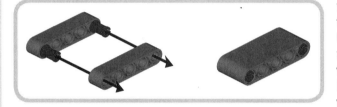

Figure 4-18: Two friction pegs connecting two 5M beams

There are two main types of pegs: smooth pegs and friction pegs. *Smooth pegs* can swivel freely in place; *friction pegs* cannot. Friction pegs stiffly keep their position, but not so stiffly as to be immovable. The NXT set includes six different types of friction pegs (numbered 14 through 19 in Figure 4-13) and four different types of smooth pegs (numbered 20 through 23 in Figure 4-13). In terms of quantity, the NXT set includes mostly friction pegs, since you'll use these most often. If you build the example shown in Figure 4-19—using a 1 × 6 TECHNIC brick, a friction peg, and a peg—you'll better understand the concept of friction pegs and smooth pegs. Just twist the pegs with your fingers.

Figure 4-19: Twist each peg to feel the difference between a friction peg and a smooth peg.

NOTE We use the term *smooth peg* when referring to smooth pegs in general (i.e., the smooth peg subcategory), but we drop the word *smooth* when referring to a specific smooth peg. When referring to a specific friction peg, however, I will always include the word *friction* in its name.

Do we measure pegs? In most cases, no. We can correctly identify most pegs by their names alone. However, the two most basic pegs—the peg and the friction peg—each have a slightly longer counterpart that we'll designate the 3M peg and the 3M friction peg, respectively. Do you remember what the actual length of a module is? It's about 8 mm. Not by accident, the friction peg and peg are 16 mm long, which corresponds to two modules (2M). The longer peg and longer friction peg are each just 8 mm longer, for a total of 24 mm (3M).

NOTE In the NXT set and other more recent LEGO sets, pegs are color-coded to help you distinguish between friction pegs and smooth pegs. All smooth pegs are tan or medium stone gray; all friction pegs are black or blue.

Finally, the NXT set includes two types of pieces that are called *peg accessories*: the steering link and the 9M steering link (Figure 4-20). (If you compare the black 9M steering link to a 9M beam, you can prove to yourself

---

[*] Pegs are also commonly known as *pins*.

that they are the same length.) These pieces are comple-
ments to the axle ball peg and the friction ball peg, and they
offer very flexible forms of movement. If you want to better
understand the flexibility of steering links, build either of
those pictured in Figure 4-20. Incidentally, the famous Alpha
Rex robot that appears on the front cover of the NXT set
uses steering links.

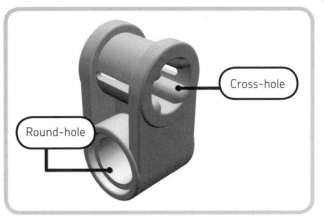

Figure 4-21: The cross block

Figure 4-20: Steering links and ball pegs work together. Both types of ball
pegs work with both types of steering links.

the connector blocks

Connector blocks are unique in that they are connectors in
every sense and rightfully belong in the connectors category,
but they usually require that you use them with pegs, axles,
or both—which are connectors themselves! If you briefly
glance back at Figure 4-13 and the pieces numbered 26
through 43, you'll get a sense for the diversity of this subcategory. Measuring is unnecessary for most of these pieces.
Figure 4-21 shows the cross block, a very common and
useful connector block, which has both a round-hole and a
cross-hole.

The purpose of the cross block or any other connector
block is to enhance your construction abilities. Although you
could create robots entirely out of beams, pegs, and axles,
connector blocks help you build more interesting and complex structures and mechanisms. For example, Figure 4-22
shows how two cross blocks—in combination with an axle, a
bushing, and friction pegs—can position a beam in a manner
that would be difficult to achieve using just beams with pegs
or axles. The projects in Part IV will show you many ways to
creatively and effectively employ connector blocks.

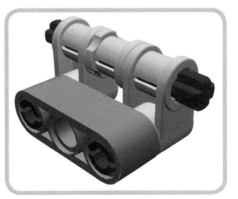

Figure 4-22: Connector blocks help you to create more
interesting and complex structures and mechanisms.

the gears

Except for the electronic elements, LEGO gears are probably the most fascinating pieces in the NXT set. The term *gear* encompasses a variety of pieces that transmit motion. Since a gear generally fits the description of a wheel with teeth, it's sometimes called a *gearwheel*. Figure 4-23 presents the various gears included in the NXT set; match up the numbers above the pieces with the numbers in Table 4-3 for information about each piece.

table 4-3: the NXT gears

number in figure 4-23	piece name	piece color (in NXT set)
1	8t (spur) gear	Medium stone gray
2	16t (spur) gear	Medium stone gray
3	24t (spur) gear	Medium stone gray
4	40t (spur) gear	Medium stone gray
5	12t double bevel gear	Black
6	20t double bevel gear	Medium stone gray
7	36t double bevel gear	Black
8	Worm gear	Black
9	Knob wheel	Black
10	Turntable	Black/dark stone gray

Figure 4-23: The gears in the NXT set

How do LEGO gears transmit motion? They accomplish this task through their teeth, as Figure 4-24 demonstrates. When the teeth of two gears *engage* or *mesh*, the rotation of any one gear causes the other gear to rotate. Notice that the gears are mounted on axles by means of their cross-holes, and the axles are mounted in round-holes so that they can rotate freely. We set up most LEGO gears in this manner. As you learned earlier, motion originates with the axles, and gears generally transmit motion between axles. Watching LEGO gears in action is exciting, but so is building with them—especially when you learn to utilize the underlying properties that govern their operation. We'll cover gearing techniques in Chapter 6.

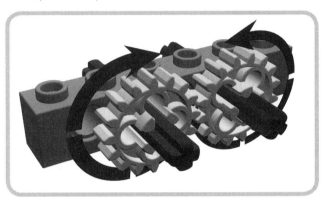

Figure 4-24: When two gears mesh, motion can transfer between the gears and, subsequently, their axles.

At this point you might be wondering, "How do we measure gears?" The answer is simple: We count teeth! With most gears we can simply count the number of their teeth and then abbreviate *teeth* with the letter *t*. For example, a gear with 16 teeth would have a measurement of *16t*. The gears in Figure 4-24 are 16t gears.

The LEGO Group has introduced a variety of gears over the years, but there are just three subcategories of gears in the NXT set. You can combine a gear's subcategory name with its measurement to get its complete name. Let's observe these three subcategories:

* Spur gears
* Double bevel gears
* Other gears

spur gears

A *spur gear* is the simplest and most common kind of gear (Figure 4-25), and it typically engages other gears positioned on parallel (non-intersecting) axles, as shown in Figure 4-24. There are four types of spur gears in the NXT set: the 8t, 16t, 24t, and 40t. Due to the LEGO spur gear's prevalence, we usually omit the *spur* before its name: For example, we call the spur gear with twenty-four teeth a *24t gear*, not a *24t spur gear*.

Figure 4-25: The 8t gear

double bevel gears

A *double bevel gear* (Figure 4-26) is a truly unique piece. Its uniqueness lies in the fact that it can use its specially-shaped teeth to act like two different types of gears. First, like bevel gears, double bevel gears can mesh when positioned on axles that are *not* parallel (*skewed*), usually engaging at perpendicular angles (Figure 4-27). Second, like spur gears, double bevel gears can mesh when positioned on axles that *are* parallel, as shown in Figure 4-24. The NXT set doesn't contain any bevel gears, but it contains three kinds of double bevel gears: the 12t, 20t, and 36t. (Note that, when referring to these pieces, we do keep the *double bevel* before their names.)

Figure 4-26: The 20t double bevel gear

Figure 4-27: Here 20t double bevel gears are engaging on skewed axles.

other gears

I've included the final three types of gears in this category: the worm gear, the knob wheel, and the turntable. For various reasons, we don't include measurements as part of their names.

The *worm gear* can engage all the toothed gears in the NXT set, but only when they're positioned on skewed axles. Figure 4-28 illustrates how a worm gear meshes with another gear, in this case, a 24t gear. While rotating, the worm gear's teeth (or *tooth* as you'll learn in Chapter 6) slide across the other gear's teeth to make the other gear rotate. Interestingly, the worm gear must *always* turn the other gear, which the arrow in Figure 4-28 signifies. Some of the main reasons for using the worm gear are that it can create significant *torque* (power) and it can greatly reduce rotational speed. We'll explore these concepts in Chapter 6.

Figure 4-28: In this setup, a worm gear engages a 24t gear. The worm gear always rotates the other gear, not vice versa.

While classifying the *knob wheel* as a gear is a bit of a stretch, I've done so because it functions as a gear: It transmits motion from one axle to another. This piece, however, has the limitation of only working with another knob wheel.

In other words, the knobs on two knob wheels "mesh," causing the same rotary motion produced by the meshing of toothed gears. On the other hand, an advantage of knob wheels is that they mesh equally well when positioned on parallel and skewed axles.

The *turntable* is a powerful gear made up of two parts, one dark stone gray and the other black, that can turn independently; twist both parts with your hand to see for yourself how this functions. If you look carefully at your turntable, you'll notice that there are teeth on the outside of the black part and the inside of the dark stone gray part. Thus, if you secure one part to a robot and engage the other part's teeth with another gear, the engaged part moves or rotates. How is this useful? You can build an entire section of a robot on the engaged part, which allows you to easily (and safely!) rotate part of your robot. Robotic arms and other stationary creations often use the turntable in this manner.

the miscellaneous elements

We have reached the fifth and final category of NXT pieces, the miscellaneous elements. Defining these pieces is rather simple: A miscellaneous element is a piece that does not fit into any of the previous four categories. Figure 4-29 presents all of these miscellaneous pieces; match up the numbers by the pieces with the numbers in Table 4-4 for information about each piece. Although these pieces may not seem especially useful, you'll find that they can be indispensable in some of your projects.

The applications of these pieces are as varied as the pieces themselves. The blue and red balls are items your robot can use in an activity; you might design a robot to find the ball, push the ball, collect the ball, throw the ball, kick the ball, or do something else with the ball (be creative!). The TECHNIC pincer, TECHNIC tooth, and 1 × 1 cone are great decorative pieces, but they can also serve important functions (the pincer and tooth can be especially useful in grabbing mechanisms). In the NXT set the balloon wheel, balloon tire, and pulley wheel typically act as robot wheels, but even these pieces can serve unconventional purposes. In Part IV you'll see some of the many ways these miscellaneous pieces can be put to use.

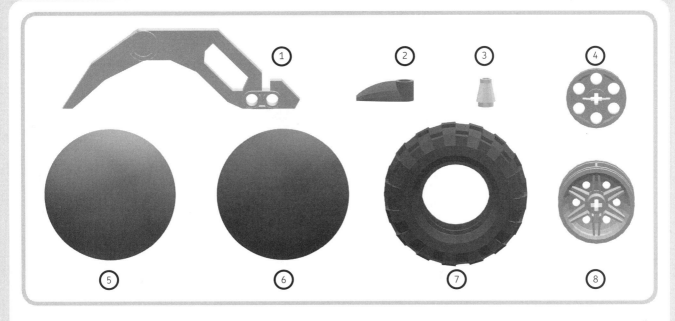

Figure 4-29: The miscellaneous elements in the NXT set

table 4-4: the miscellaneous NXT elements

number in figure 4-29	piece name	piece color (in NXT set)
1	TECHNIC pincer	Pearl gray
2	TECHNIC tooth	Orange
3	1 × 1 cone	White
4	Medium pulley wheel	Medium stone gray
5	Ball	Blue
6	Ball	Red
7	Balloon tire	Black
8	Balloon wheel	Medium stone gray

conclusion

Exploring the topic of LEGO MINDSTORMS NXT construction is incredibly fun, but the magnitude of the subject is astonishing. As you learned in this chapter, one of the first steps is to acquire a solid understanding of the LEGO pieces in the NXT set. We began by familiarizing ourselves with the NXT pieces as a whole and detailing some underlying concepts, and then we proceeded to examine each of the five categories of pieces in the NXT set: electronics, beams, connectors, gears, and miscellaneous elements. In the following chapter, you'll learn practical techniques for building effective structures for NXT robots.

building sturdy structures

The overall robustness or sturdiness of an NXT robot depends mainly on its structure, which is the basis of its design. And having a strong structure is important because a flimsy NXT robot is liable to simply fall apart. As you learned in Chapter 4, beams are the main structural pieces in the NXT set; hence, building sturdy structures begins with learning to use beams effectively.

In this chapter we'll examine some basic building techniques for using beams to make structures: extending beams, widening beams, forming corners, and creating angled, dynamic, and flexible structures. These fundamental techniques offer viable solutions for a variety of NXT creations, but they can also serve as the foundation for more complex techniques, some of which you'll see when we build complete robots in Part IV.

I encourage you to pull out your NXT set and build the examples in this chapter as you go along.

NOTE I will not be covering construction techniques specific to TECHNIC bricks because the NXT set relies almost entirely on the use of straight beams and angled beams. If you'd like to learn more about TECHNIC bricks, consult Appendix C for related Internet resources.

extending beams

Most of the beams in the NXT set only come in odd sizes ranging from 3M to 15M, so what should you do when you need a beam with an even module measurement, a length exceeding 15M, or both? Fortunately, you can easily extend a beam to the exact desired size.

Figure 5-1 demonstrates a robust technique for extending beams. In this example, an 11M beam overlaps a 15M beam, connecting with two friction pegs to achieve a total length of 23M. To achieve greater lengths, you could simply use a third beam as shown in Figure 5-2. In fact, you can continue attaching beams in this way to keep on extending them. If you need stronger extended beams, you can use 3M friction pegs and a fourth beam as shown in Figure 5-3.

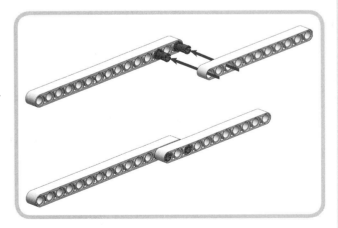

Figure 5-1: Extending a 15M beam with an 11M beam and two friction pegs to achieve a length of 23M

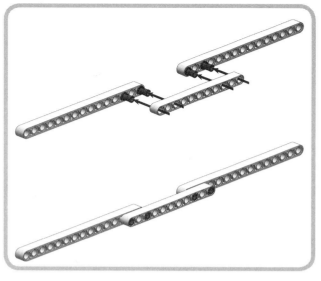

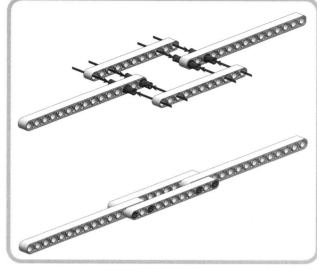

Figure 5-2: Repeating the technique shown in Figure 5-1 extends the structure to a length of 35M.

Figure 5-3: This structure, based on the one in Figure 5-2, uses 3M friction pegs and an additional 11M beam for greater strength.

NOTE In Figures 5-1 through 5-3, shifting a beam by one round-hole (one module) would produce an even module measurement. For example, shifting a 15M beam inward by one round-hole in Figure 5-2 would result in a total length of 34M.

Furthermore, as the structures in Figures 5-1 through 5-3 demonstrate, when extending beams, you should use friction pegs instead of smooth pegs because they have a stronger grip, and you should generally maintain a minimum overlap of 3M. Overlapping is a particularly important construction principle, as more overlap creates stronger structures. An overlap of 3M is good for many purposes, but if an extended beam will be supporting significant weight or otherwise encountering substantial force, you should increase the overlap to 4M or greater. You can also achieve more strength by using more pegs. For example, when connecting the beams in Figure 5-3, we could have used six 3M friction pegs instead of four. If you need a much stronger extended beam, consider increasing both the overlap *and* the number of pegs used.

NOTE Remember that we use the term *smooth peg* when referring to smooth pegs in general (i.e., the smooth peg subcategory), but we drop the word *smooth* when referring to a specific smooth peg.

widening beams

A single beam is relatively thin—only 8 mm. Most structures, however, demand much greater widths. An easy way to address this problem is by widening beams. Figure 5-4 shows an example of this technique, which once again uses friction pegs for an easy and sturdy solution. If you need a structure that is three beams wide, 3M friction pegs work well (Figure 5-5). Another easy way to achieve a three-beam width is to use the 3M pegged block with two beams (Figure 5-6). The pegs on the 3M pegged block are more like smooth pegs than friction pegs, but they still work well.

Figure 5-4: Widening a section of a 13M beam to a two-beam width using a 7M beam and two friction pegs

Figure 5-5: Widening a section of a 13M beam to a three-beam width using two 7M beams and two 3M friction pegs

Notice that the examples in Figures 5-4 and 5-5 widen only part of a beam, whereas the example in Figure 5-6 widens an entire beam. Just like you can extend beams in a variety of ways, you can widen beams in a variety of ways. Note also that the beams in Figures 5-4 and 5-5 have an overlap of 7M, whereas the beams in Figure 5-6 overlap the 3M pegged block by 3M.

Figure 5-6: Widening a 5M beam to a three-beam width using another 5M beam and a 3M pegged block

forming corners

For the NXT set, the LEGO Group developed a new piece called the *5M pegged perpendicular block*, which can be used with beams to form right angles. This technique is useful in a variety of creations. For example, many stationary robots that sit on a flat surface can benefit from a square base (which can be created by using beams and 5M pegged perpendicular blocks to form four corners). Figure 5-7 shows how to create a right angle with two 11M beams and the 5M pegged perpendicular block.

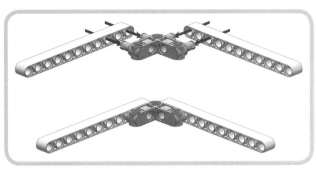

Figure 5-7: The 5M pegged perpendicular block can be used with beams to form corners.

creating angled structures

We can use straight beams to create many useful structures, but using angled beams or a combination of straight beams *and* angled beams enables us to produce more advanced structures. When building angled structures, the techniques we've discussed so far are still valid. However, we face a compatibility issue: Some angled beams have a few cross-holes, but most of the other holes involved are round-holes. How do we effectively connect a round-hole to a cross-hole for structural purposes?

The answer lies in the use of friction axle pegs and axles. The friction axle peg is a blue peg with an axle on one end and a friction peg on the other, which allows one end to firmly grip a round-hole and the other to firmly grip a cross-hole. Figure 5-8 illustrates how to widen a 9M angled beam using another 9M angled beam and two friction axle pegs. We could have used friction pegs to connect these two beams, but they wouldn't have resulted in as much overlap.

Figure 5-8: Widening a 9M angled beam with a 9M angled beam and two friction axle pegs

Figures 5-9 and 5-10 illustrate techniques involving a combination of straight beams and angled beams. In Figure 5-9, we extend a 9M beam using a 7M angled beam, a friction peg (round-hole to round-hole), and a friction axle peg (round-hole to cross-hole). Figure 5-10 shows how to widen a beam with two 5M perpendicular angled beams joined by a 3M friction peg and a 3M axle. Notice how the 3M axle connects all three pieces by passing through two cross-holes and a round-hole. Note also the 4M overlap in both techniques that adds robustness to the connections.

Figure 5-9: Extending a 9M beam with a 7M angled beam using a friction peg and a friction axle peg

Figure 5-10: Widening a 9M beam with two 5M perpendicular angled beams using a 3M axle and a 3M friction peg

creating dynamic structures

All the previous techniques focused on creating rigid structures, but *dynamic structures* are characterized by continuous motion. We typically integrate this type of structure with gears to transform simple rotary motion into more complex forms of motion. (We'll be looking at gears more closely in Chapter 6.) Given their nature, dynamic structures often use smooth pegs.

Let's build a basic dynamic structure that transforms the rotary motion of a gear into a swinging motion commonly used in walking creations. Figure 5-11 shows the parts you'll need, and Figure 5-12 presents the building instructions. Follow the three construction steps, using an 11M beam, two 9M beams, a 40t gear, a 4M axle, a bushing, two pegs, and a 3M peg. When you've built the structure, hold the 11M beam in one hand and rotate the axle with your other hand. The 9M beam attached to the 40t gear begins swinging up and down. With further development, you could turn this into a useful robot leg!

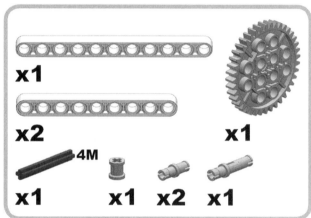

Figure 5-11: You'll need these parts to make the dynamic structure in Figure 5-12.

creating flexible structures

Flexible structures can expand and contract, relying on smooth pegs (as dynamic structures do) or even axles to achieve their flexibility. A folding chair is an excellent real-world example of a flexible structure: You can open it when you want to use it, you can close it when you're finished, and you can even put it in a position somewhere in between fully open and fully closed (an *uncomfortable* position).

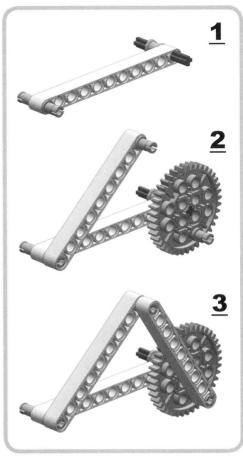

Figure 5-12: This dynamic structure can continuously produce a swinging motion suitable for the legs of a walking robot.

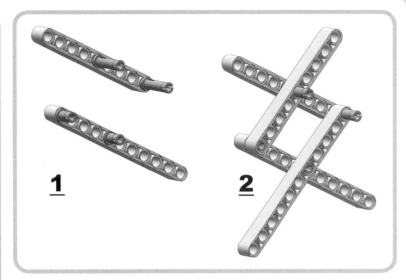

Figure 5-13: This flexible structure can expand or contract (shown expanded).

Figure 5-13 shows how to make a simple flexible LEGO structure. First, insert two 3M pegs into a 9M beam (the beam on top) and insert two pegs into an 11M beam (the beam on the bottom). Next, connect a 9M beam to the two leftmost pegs and connect an 11M beam to the two rightmost pegs.

NOTE I would normally use all pegs in the dynamic and flexible structures shown here, instead of including some 3M pegs. (In this case, the 3M pegs cause the beams to slip a bit.) However, because of the limited number of pegs in the NXT set, I substituted the 3M pegs.

If you play with this structure, you'll see how it can swiftly expand into a long, thin structure or contract into a compact pile of beams. The official NXT robot Spike, a robotic scorpion, employs a more complex version of this structure as its tail, using a motor to expand and contract it in order to "sting." You could also integrate flexible structures with sensors. For example, a flexible structure could serve as a sort of lever or button that activates a sensor when the user changes the structure's form (i.e., pushes or pulls it). It's not necessary that you integrate flexible structures with motors or sensors, but NXT robots can generally utilize flexible structures more effectively when you do.

conclusion

In this chapter, we've examined some basic techniques for building structures using pieces from the NXT set, specifically beams. You learned how to extend and widen beams in several different ways, form corners with the 5M pegged perpendicular block, create angled structures using angled beams or a combination of angled beams and straight beams, and create dynamic structures and flexible structures which achieve their flexibility through pegs and axles. In the next chapter, we'll conclude Part II with an in-depth discussion of LEGO gears.

building with gears

In Chapter 4 we briefly discussed the gears in the NXT set. We noted that their general purpose is to transmit motion, and we outlined three subcategories: spur gears, double bevel gears, and "other" gears. In this chapter we'll expand our discussion of the gears and learn how to *effectively* transmit motion through a variety of gearing techniques. We'll begin with a key gearing concept, the gear train, and establish some important terminology. Next, we'll explore how to properly control the performance and manage the assembly of systems of gears. Finally, we'll take a closer look at the worm gear and the turntable.

the gear train

Whenever we use two or more gears to transmit motion, we call the series of gears a *gear train*. To help us understand some of the concepts surrounding the gear train, let's build a simple mechanism with a crank that can turn three gears. Figure 6-1 shows the parts you'll need: two 1 × 6 TECHNIC bricks, one 5M perpendicular angled beam, one 4M axle, one 20t double bevel gear, two 12t double bevel gears, one 3M peg, two axle pegs, two half-bushings, and a peg extender. Figure 6-2 presents the building instructions.

Figure 6-1: You'll need these parts to make the mechanism in Figure 6-2.

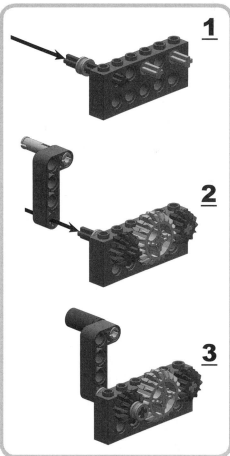

Figure 6-2: Building a simple mechanism with three gears and a crank

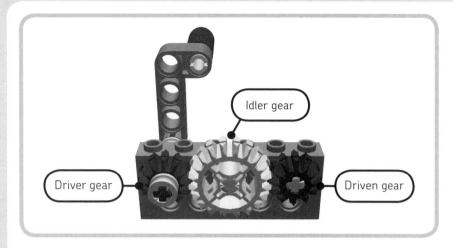

Figure 6-3: A gear can serve one of three roles in a gear train: the driver gear, the driven gear, or an idler gear.

Within a gear train, a gear can serve one of three roles: the driver gear, the driven gear, or an idler gear. Figure 6-3 shows which gears perform which roles in the mechanism you just built. The *driver gear* is the first gear in the gear train or, in other words, the first gear to transmit motion. The *driven gear* is the last gear in the gear train or, in other words, the last gear to transmit motion.

Finally, an *idler gear* serves two purposes. The first purpose is to transmit motion from the driver gear to the driven gear when they're too far apart for their teeth to mesh. The second purpose is to reverse the direction of rotation. A fundamental property of gears is that any two engaged gears revolve in opposite directions. Hence, we can control whether a driven gear rotates clockwise or counterclockwise by including an idler gear in the gear train.

NOTE Although you can often successfully mount idler gears on axle pegs, it is more common to mount driven gears on axles.

If you turn the crank on the mechanism, you can see the 20t double bevel idler gear perform both of its functions. First, it transmits motion from the 12t double bevel driver gear to the 12t double bevel driven gear. Second, it reverses the direction of rotation, causing the driven gear to revolve in the same direction as the driver gear. If the driver gear revolves clockwise, for example, the idler gear revolves counterclockwise, and then the driven gear revolves clockwise.

NOTE In order for the driver gear and the driven gear to revolve in the same direction, the gear train must consist of an odd number of gears. Compound gear trains, which we'll cover in "Compound Gearing: Achieving Greater Torque or Speed" on page 56, are an exception to this rule.

This gear train has only one idler gear, but you can theoretically have as many idler gears in a gear train as you want—or none at all. More gears introduce more friction, which is *not* desirable. Thus, where possible, having fewer gears is better than having more gears.

controlling a gear train's performance

Not every gear train performs the same. When referring to a gear train's performance, we're talking about the speed and torque of the driven gear's axle. In simplified terms, *speed* is how fast the axle revolves, and *torque* is the strength or power with which the axle revolves. Using several techniques—gearing down, gearing up, and compound gearing—we can control a gear train's performance, adjusting it for specific purposes. Before exploring these topics, however, we must discuss gear ratios.

the gear ratio

We can measure or describe the performance of a gear train with a *gear ratio*, which is the number of rotations of the driver gear relative to the number of rotations of the driven gear. For example, a gear ratio of 1:3 (pronounced *one to three*) indicates that for every rotation of the driver gear, the driven gear rotates three times. You'll learn how gear ratios relate to speed and torque in a moment when we discuss the techniques for controlling a gear train's performance. Let's first concentrate on how to calculate gear ratios themselves, beginning with two basic gearing principles.

First, each tooth on a gear can turn exactly one tooth on another gear. If a gear train consists of two gears, each with the same number of teeth, the gear ratio for that gear train would be 1:1. That is, for every rotation of the driver gear, the driven gear also completes one rotation. Therefore, by comparing the number of teeth on the driver gear to the number of teeth on the driven gear, we can determine how many rotations the gears will complete relative to each other.

Second, idler gears don't affect performance except in regard to friction. We ignore any idler gears in a gear train when determining a gear ratio. The gear ratio of the mechanism we built earlier would be 1:1 because the driver gear and driven gear have the same number of teeth and the 20t double bevel gear is only an idler gear.

With these concepts in mind, we can follow a simple three-step method to determine the gear ratio of most LEGO gear trains. Let's quickly modify our mechanism and then use this method to determine its gear ratio. Remove the driven gear and its axle peg so that the mechanism looks like the one in Figure 6-4. The 12t double bevel gear is still the driver gear, but now the 20t double bevel gear is the driven gear. Follow these steps to determine the gear ratio:

1. **Determine the number of teeth on the driver gear and driven gear.** In our example, the driver gear has 12 teeth and the driven gear has 20 teeth.

2. **Form a ratio with the driven gear's teeth first and the driver gear's teeth second.** Continuing our example, we would have a ratio of 20:12.

3. **Simplify the ratio to its lowest terms.** In this case, 20:12 simplifies to 5:3. That is, for every five rotations of the driver gear, the driven gear rotates three times.

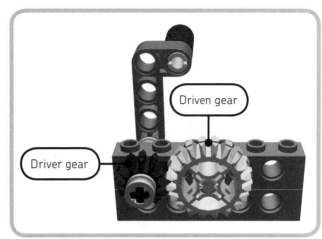

Figure 6-4: A 12t double bevel gear driving a 20t double bevel gear results in a gear ratio of 5:3.

gearing down: more torque and less speed

Gearing down a gear train increases its torque but decreases its speed. We can gear down by driving a larger gear with a smaller one. When we gear down, the resulting gear ratio will always consist of a larger number followed by a smaller number because the driver gear completes more rotations than the driven gear (e.g., 5:1, 3:1, or 5:3). The gear train in Figure 6-5 shows an 8t gear driving a 40t gear. By following our three-step method, we are able to determine that the gear ratio is 5:1. This ratio tells us that the driver gear rotates five times for every one rotation of the driven gear; speed *decreases* by a factor of five, but torque *increases* by a factor of five.

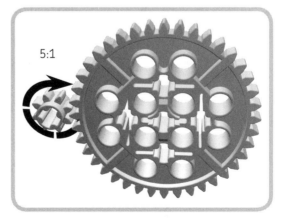

Figure 6-5: Gearing down—driving a larger gear with a smaller one—produces more torque but less speed.

gearing up: more speed and less torque

Gearing up a gear train increases its speed but decreases its torque. We can gear up by driving a smaller gear with a larger one. When we gear up, the resulting gear ratio will always consist of a smaller number followed by a larger number because the driver gear has more teeth and completes fewer rotations than the driven gear (e.g., 1:5, 1:3, or 3:5). The gear train in Figure 6-6 shows a 40t gear driving an 8t gear, which produces a gear ratio of 1:5. This ratio tells us that the driver gear completes one rotation for every five rotations of the driven gear; speed *increases* by a factor of five, but torque *decreases* by a factor of five.

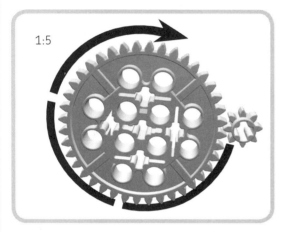

Figure 6-6: Gearing up—driving a smaller gear with a larger one—produces more speed but less torque.

compound gearing: achieving greater torque or speed

We just saw that an 8t gear and a 40t gear can produce ratios of either 1:5 or 5:1. What if you want to achieve gear ratios that provide even *more* torque or speed? The worm gear and turntable (covered later in this chapter) can provide larger gear ratios, but we won't always want to use these specialized gears. Alternatively, we can implement a powerful technique known as *compound gearing*, which places two gears on the same axle. We call a gear train that uses this technique a *compound gear train*. Figure 6-7 shows a compound gear train with four gears: two 8t gears and two 24t gears. A 24t gear and an 8t gear share an axle in the middle of the gear train. What is the gear ratio of this compound gear train? Let's calculate it.

Calculating the gear ratio of a compound gear train is a bit different than calculating the gear ratio of a regular gear train. Follow these steps:

1. **Determine the number of teeth on the driver gear, the driven gear, and any gears that share an axle.** In our example, we have to count the teeth on all the gears. The driver gear has 8 teeth, the driven gear has 24 teeth, and the two gears that share an axle in the middle of the gear train have 8 teeth and 24 teeth, respectively.

2. **Calculate the gear ratios.** Continuing our example, the first pair of gears is an 8t gear driving a 24t gear, which

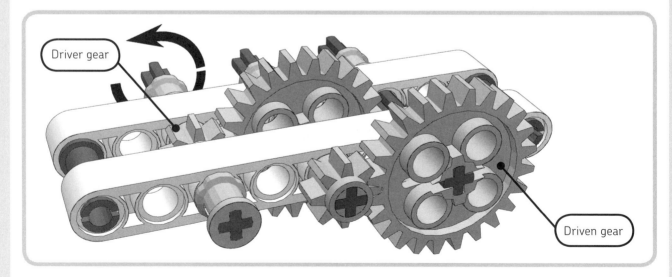

Figure 6-7: Compound gearing places two gears on the same axle, and in this figure a 24t gear and an 8t gear share the middle axle of the gear train.

results in a ratio of 3:1. The next pair of gears is also an 8t gear driving a 24t gear, which means it also results in a gear ratio of 3:1.

NOTE When calculating the gear ratio for compound gear trains, we ignore all idler gears that do *not* share an axle.

3. **Multiply the gear ratios like fractions.** In this case, multiplying 3:1 and 3:1 produces a gear ratio of 9:1.

4. **Simplify the ratio to its lowest terms.** We don't need to simplify the gear ratio in this example because it's already in lowest terms.

For every nine rotations of the driver gear, the driven gear rotates once; the speed *decreases* by a factor of nine, but the torque *increases* by a factor of nine. If the 24t gear were the driver gear instead of the driven gear, the gear ratio would be 1:9. This is a simple example of compound gearing—it's possible to gear up or down much more significantly by using multiple stages of compound gearing.

achieving optimal performance

With all of these options for controlling gear train performance, you may be wondering when you should gear up, gear down, or do neither (i.e., use a gear ratio of 1:1). The answer depends on what purpose the gear train serves. For example, if the gear train is part of a mechanism that works with relatively heavy objects, gearing down for more torque would probably be the best choice. If it's not immediately obvious how you should configure a gear train's performance, experimenting with different gear ratios will help you decide.

If you've chosen to gear up or down, your next question would be, "How much?" As you work with the NXT servo motors and different gear trains, you'll develop a sense for the strengths and speeds of various gear ratios, and you'll soon know which ratios are best for which projects. Nevertheless, in many cases, further experimentation will still be the solution.

NOTE As a general rule, don't gear down simply for the sake of decreasing speed. If you need to decrease speed but don't necessarily need to increase torque, you can use programming to decrease the speed of the motor powering the gear train. You'll learn how to do this in Chapter 7, the first programming chapter.

Keep in mind that you'll usually have to make a compromise between speed and torque. For example, if you're creating a racecar-style NXT robot, you would want the robot to go as fast as possible. Does this mean you should gear up enormously to achieve maximum speed? Not exactly. If you gear up too highly, the torque will decrease so drastically that there won't be enough power to move the car itself! In this case, it would be better to experiment and then settle on a gear ratio that gives the robot sufficient speed while still giving it enough torque to efficiently move.

assembling LEGO gear trains

The primary challenge when assembling a gear train—apart from deciding which gears to use—is in properly spacing each pair of gears. If the gears are too far apart or too close together, their teeth can't mesh and therefore can't transmit motion. How you should space the gears generally depends on whether they are on parallel axles or perpendicular axles.

spacing gears on parallel axles

The mechanism we built earlier exhibits the most basic type of gear train, which consists of parallel gears mounted in the same row of round-holes. With this setup, spacing is in increments of 1M because the round-holes on beams are separated by a distance of 1M. But how many round-holes apart should we space a given pair of gears? The module itself provides the answer.[*]

---

[*] There are beams that have round-holes spaced differently than the round-holes on normal beams. The NXT set doesn't include any of these special beams, however.

If we use the module to measure the radius of each of the spur gears and the double bevel gears in the NXT set—the *radius* being the distance from the center of a gear to the outermost part of its teeth—we'll gather the data presented in Table 6-1. (I have excluded the "other" gears in the NXT set because of differences in their design.) Adding the radii of any two gears tells us how far apart the gears' axles must be in order for the gears' teeth to mesh. The sum of the radii must be a whole number (1, 2, 3, and so on) because the round-holes are 1M apart.

table 6-1: radii of spur gears and double bevel gears in the NXT set

gear name	radius
8t gear	0.5M
16t gear	1M
24t gear	1.5M
40t gear	2.5M
12t double bevel gear	0.75M
20t double bevel gear	1.25M
36t double bevel gear	2.25M

As shown in Table 6-1, a 12t double bevel gear has a radius of 0.75M and a 20t double bevel gear has a radius of 1.25M. Now we know why each pair of gears meshes perfectly in the mechanism we built at the beginning of this chapter: The sum of their radii is 2M, which is a whole number.

Let's try another example. If you wanted to use a 40t gear (which has a radius of 2.5M) and a 24t gear (which has a radius of 1.5M), would this combination of gears work? Add up the radii of these gears, and you'll get 4M. This is a whole number, so it will indeed work (Figure 6-8). How about using a 16t gear (1M) and an 8t gear (0.5M)? Add up the radii of these gears, and you'll get 1.5M. This combination of gears *won't* work because the sum of their radii is not a whole number.

How should you space knob wheels? Remember that you can only use a knob wheel with other knob wheels. When positioned on parallel axles, knob wheels should be spaced by 2M (Figure 6-9).

(Front view)

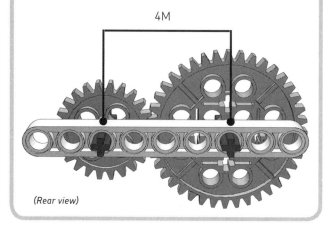

(Rear view)

Figure 6-8: A 24t gear has a radius of 1.5M and a 40t gear has a radius of 2.5M, so we must space the gears' axles by 4M, the sum of the radii.

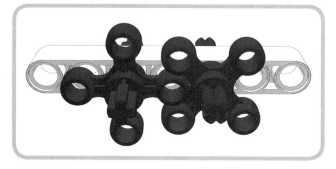

Figure 6-9: These two knob wheels demonstrate that spacing should be 2M when they are on parallel axles.

WORKING WITH IMPERFECT GEAR COMBINATIONS

If you attempt to use a spur gear and a double bevel gear, their radii total a mixed number (e.g., 1.75). Moreover, adding the radii of some spur gear combinations (such as a 16t gear and a 24t gear) and some double bevel gear combinations (such as a 20t double bevel gear and a 36t double bevel gear) also results in mixed numbers. When working on parallel axles, is it possible to use these "imperfect" gear combinations? Yes, it can be done. The key is to find creative ways to space gears in increments other than 1M.

Figure 6-10 illustrates five different imperfect gear combinations. In each combination, the gears are positioned in round-holes on different levels to approximate the necessary spacing. Because imperfect gear combinations are less efficient and more difficult to work with than "perfect" gear combinations, you should generally use the latter whenever you can.

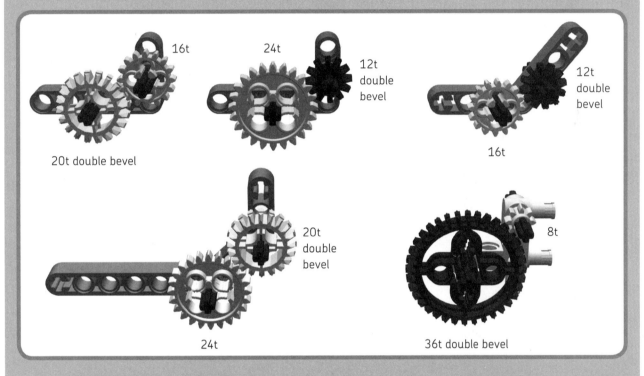

Figure 6-10: Five "imperfect" gear combinations that work when the gears are positioned on round-holes at different levels.

spacing gears on perpendicular axles

We can also use the double bevel gears or the knob wheels on perpendicular axles to transmit motion at right angles. With this type of gear train, we don't add up the gears' radii and try to calculate the necessary spacing. Instead, for any given combination of these gears, we simply experiment with the spacing until we find something that works.

There are two gear trains that mesh perfectly at a right angle, and you should memorize them. One of these involves two 20t double bevel gears (Figure 6-11), and the other involves a 20t double bevel gear and a 36t double bevel gear (Figure 6-12). To build these examples, first construct a solid structure by pushing the round-holes of a 7M perpendicular angled beam onto the studs of two TECHNIC bricks. Next,

place two 4M axles through the round-holes and attach the gears.

NOTE Two 36t double bevel gears also mesh perfectly when positioned at a right angle, but the NXT set includes only one of these gears.

For other types of gear trains that transmit motion at right angles, you may need to use half-bushings or even bushings to adjust the position of the gears. For example, in Figure 6-13 we use two half-bushings to shift 12t double bevel gears forward into a meshing position. In Figure 6-14 only one half-bushing was necessary to get a 20t double bevel gear and a 12t double bevel gear to engage. In Figure 6-15 two half-bushings were necessary to enable a pair of knob wheels to mesh.

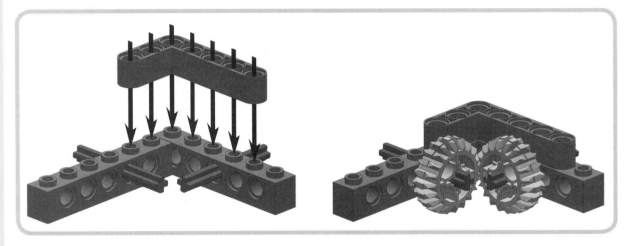

Figure 6-11: Two 20t double bevel gears engaging each other on perpendicular axles

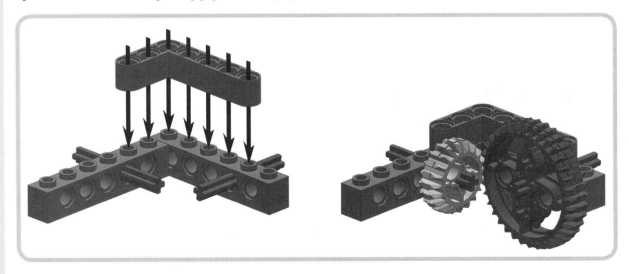

Figure 6-12: A 20t double bevel gear and a 36t double bevel gear engaging each other on perpendicular axles

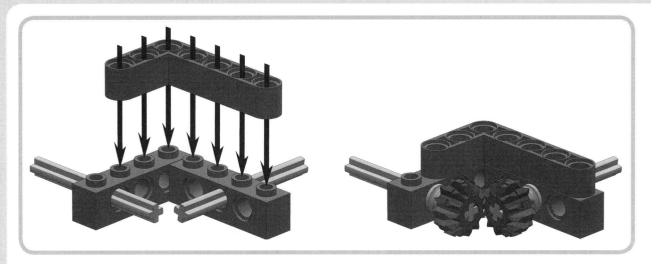

Figure 6-13: Two 12t double bevel gears engaging on perpendicular axles

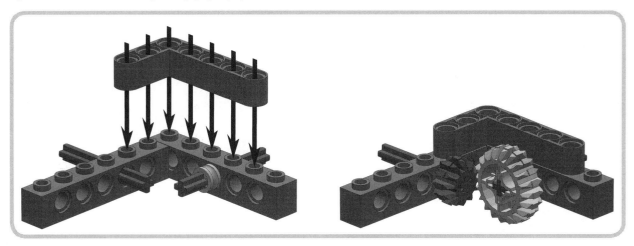

Figure 6-14: A 12t double bevel gear and a 20t double bevel gear engaging on perpendicular axles

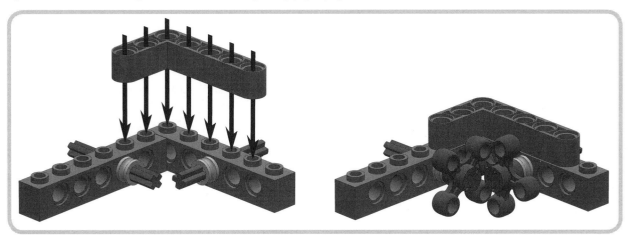

Figure 6-15: Two knob wheels engaging on perpendicular axles

the worm gear

Most people would probably consider the worm gear to be the oddest gear in the NXT set. Not only does it have an odd name, but it looks more like a screw than a gear! In fact, the worm gear is also known as the *screw gear*. However, it's to our advantage to take this gear very seriously. To better understand how the worm gear functions, let's build another mechanism with a gear train, this time featuring a worm gear and a 24t gear.

Figure 6-16 shows the pieces you'll need to build the mechanism, and Figure 6-17 presents the building instructions. The 24t gear is mounted on an axle peg, which allows it to rotate freely. The extended cross blocks use friction axle pegs to connect to the 7M beam, and the 10M axle uses two bushings and two half-bushings to keep the worm gear in place underneath the 24t gear.

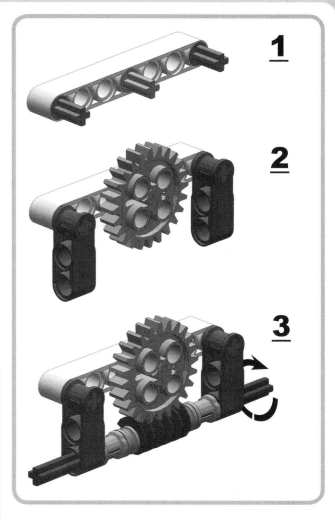

Figure 6-17: Building a simple mechanism with a gear train consisting of a worm gear and a 24t gear

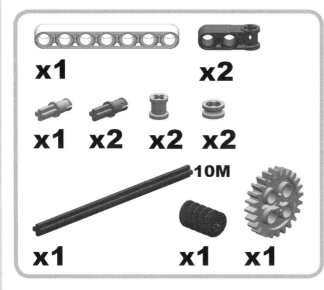

x1 x2

x1 x2 x2 x2

10M

x1 x1 x1

Figure 6-16: You'll need these pieces to build the mechanism in Figure 6-17.

From this mechanism, we can observe the three main characteristics of the worm gear:

The worm gear must always engage a gear on a perpendicular axle.

You'll find the worm gear useful in situations in which you want to transmit motion at perpendicular angles. This is much like using double bevel gears to transmit motion at perpendicular angles.

The worm gear has only one tooth.

If you twist or rotate the worm gear's axle, how fast does the 24t gear revolve? Very slowly. In fact, you'll find that for every complete rotation of the axle, the worm gear turns only one tooth on the 24t gear. The gear ratio is 24:1, so for every twenty-four rotations of the worm gear, the 24t gear rotates only once. This increases torque dramatically, but it also reduces speed dramatically. As a general rule, gear trains that use worm gears are slow but powerful.

The worm gear must always be the driver gear.

Try to turn the 24t gear, using it as the driver gear. You'll quickly see that you can't—the worm gear locks in place to prevent any movement except for its own. At first, this might seem like a limitation, but you can use this same locking feature to prevent a gear train from "unwinding" under significant pressure.

the turntable

If the worm gear is the odd gear in the NXT set, the turntable is the defiant gear. Because its teeth are on its side, the turntable cannot as easily engage other gears—except for the worm gear. In Figure 6-18 you can see how naturally a worm gear drives a turntable. (Normally, there would be additional pieces to hold everything in place.) The arrow around the worm gear's axle reminds us that the worm gear must always be the driver gear, and the arrow around the turntable shows us how the top part rotates, assuming the bottom part is secured to other pieces.

What would the gear ratio be for this gear train? First, we note that the worm gear has one tooth and the turntable has 56 teeth. Next, we form a ratio with the driven gear's teeth first and the driver gear's teeth second: 56:1. We cannot simplify the ratio any further, so we can conclude that for every 56 rotations of the worm gear, the turntable turns once. This produces quite a bit of torque, which is highly beneficial since we'll often build part of a robot onto the turntable.

Figure 6-18: A worm gear driving a turntable

conclusion

After reading this chapter, you should have a solid understanding of the LEGO gears in the NXT set—and perhaps a greater appreciation for them as well! We covered a number of gearing concepts, beginning with the gear train, which consists of two or more gears that transmit motion. We then moved on to controlling the performance of a gear train through techniques such as gearing up, gearing down, and compound gearing; and we also learned how to measure performance with the gear ratio. We also explored how to effectively assemble gear trains consisting of parallel gears or perpendicular gears, the two types of gear trains commonly found in NXT robots. Finally, we revisited the worm gear and turntable to more fully discuss their important aspects.

We've completed Part II, and after covering programming in Part III, you'll learn even more about construction through the projects in Part IV.

PART III:

programming

7

introduction to NXT-G

The ability of an NXT robot to perform any given task—whether following a line, throwing a ball, or sweeping the floor—isn't instinctive. We must provide the robot with specific instructions that tell it what to do; we must program the robot. In terms of the NXT set, *programming* generally involves writing a program on a home computer and then transferring it to the NXT, the "brain" of a robot, which can then *execute*, or run, the program. Programs can tell the NXT how to operate motors, read sensors, play sounds, and much more.

We must write all of our programs in a programming language, and this book focuses on the official NXT-G programming language included in the NXT software. As you learned back in Part I, NXT-G is a *graphical programming language*, one in which we program by clicking and dragging blocks of code onto the screen. (The *G* in NXT-G stands for *graphical*.) We'll briefly observe several unofficial text-based programming languages for the NXT in Chapter 9.

In this chapter, we'll discuss the basics of NXT-G and build a programming foundation for Chapter 8, which covers more advanced NXT-G material. First, we'll observe how to create and manage user profiles and NXT-G programs within the LEGO MINDSTORMS NXT software. Second, we'll examine the NXT-G interface, which plays a highly important role in the programming process. Third, we'll discuss some fundamental NXT-G concepts that are vital to successful programming. Finally, we'll learn how to use the first division of programming commands known as the Common blocks.

NOTE Chapters 7 and 8 cover NXT-G version 1.0. The latest release of NXT-G (1.1) offers a number of bug fixes and performance improvements but remains virtually the same in programming capabilities. For this reason, these chapters apply to both versions of NXT-G (1.0 and 1.1).

starting an NXT-G program

When you launch the LEGO MINDSTORMS NXT software, it displays the main screen from which you can navigate to the software's primary features (Figure 7-1). In order to access the NXT-G interface and begin using or working on a program, you must create a new program or open an existing one.

NOTE I'm assuming you've installed the software included with the NXT set on your computer and successfully established a USB or Bluetooth connection between your computer and NXT. If you've not yet done so, consult Chapter 2 for instructions.

First, we need to consider user profiles. A *user profile* keeps all of your programs and related files in one place. When you create a new program, the NXT software assigns that program to the current user profile so that whenever you attempt to open existing programs, only ones assigned to the current user profile are listed. As a general rule, you should always check the current user profile to make sure that it's the one you want to use (Figure 7-2). Initially, only a Default profile exists, which is fine if you're the only one using the software. If there were other user profiles, you could select one from the drop-down menu shown in Figure 7-2.

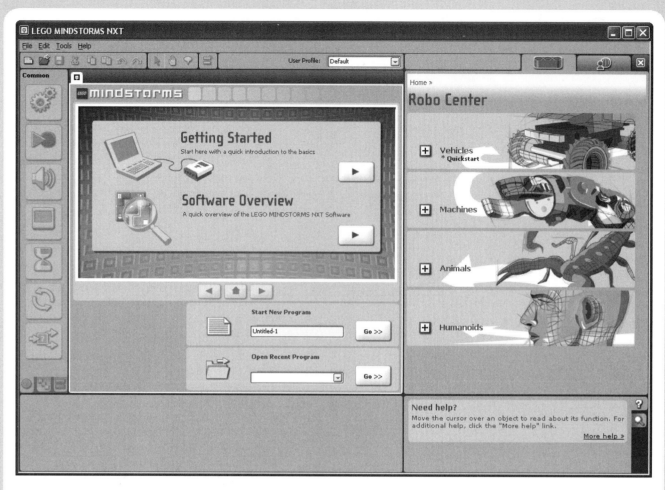

Figure 7-1: The main screen of the LEGO MINDSTORMS NXT software

Figure 7-2: You can view and change the current user profile near the top of the screen.

On the other hand, if more than one person is sharing the software, you should each consider creating your own profile so that you can keep your files separate. Selecting Edit ▶ Manage Profiles from the menu bar in the upper-left corner of the screen displays the Manage Profiles dialog (Figure 7-3). You can use the Manage Profiles dialog to create, delete, and change (rename) profiles. I've created a new user profile named *David J. Perdue* for illustrative purposes, but I ordinarily use the Default profile myself.

Figure 7-3: Use the Manage Profiles dialog to create, delete, and change user profiles.

Once you've chosen the correct profile, the next step is to open or create an NXT-G program, which you can do near the bottom of the screen (Figure 7-4). To create a new program, type a name in the text box under the words *Start New Program* and then click the associated **Go>>** button. To open an existing program that you recently accessed, simply select a program from the drop-down menu under the

words *Open Recent Program* and then click the **Go>>** button. Because there are initially no existing programs, type **Test Program** in the text box under the words *Start New Program* and click the **Go>>** button to create a new program named *Test Program*.

NOTE To access a complete list of existing programs assigned to the current user profile, select File ▸ Open or press CTRL-O (Windows) or CMD-O (Mac). You can also create a new program by selecting File ▸ New, pressing CTRL-N (Windows), or pressing CMD-N (Mac).

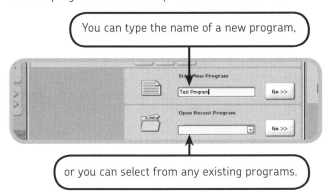

You can type the name of a new program,

or you can select from any existing programs.

Figure 7-4: Near the bottom of the screen, you can create new NXT-G programs or open recent programs.

the NXT-G interface

The NXT-G interface appears once you've created or opened a program. (The Robo Center reduces your view of the interface, so you might want to hide it as shown in Figure 7-5.

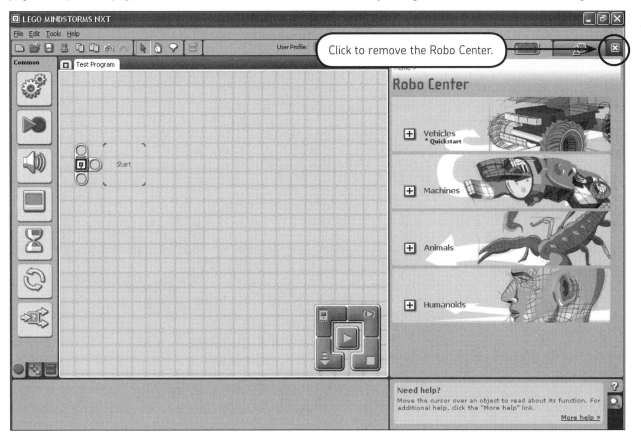

Figure 7-5: The new Test Program and the NXT-G interface

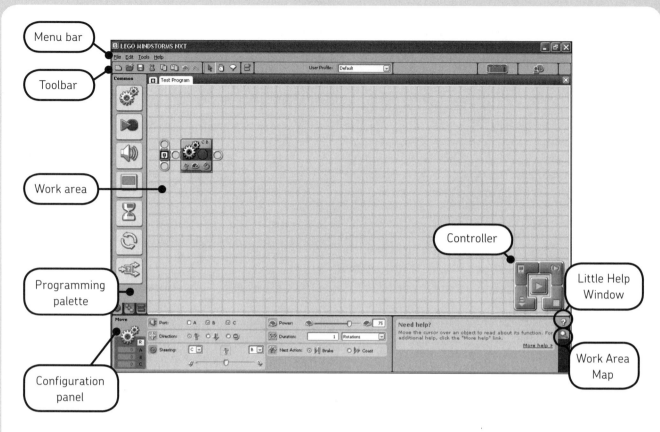

Figure 7-6: The NXT-G interface has eight main features.

You can always restore the Robo Center later by clicking the tab with the orange 3M beam.) Figure 7-6 points out the eight main features of the NXT-G interface. To activate one of the features, I added a single Move block, which we'll cover later. Let's examine each of these features in turn.

the menu bar

Positioned in the upper-left corner of the screen, the *menu bar* enables you to easily access groups of related commands through four menus: File, Edit, Tools, and Help (Figure 7-7). The File menu displays commands for creating programs, opening programs, closing programs, printing programs, and quitting the software; the Edit menu displays commands

for managing your program and its programming blocks; the Tools menu displays commands for calibrating sensors (Chapter 8 discusses calibration) and updating the firmware on an NXT; and the Help menu gives you access to documentation and other helpful resources. You'll be learning how to use the menus throughout this chapter and in Chapter 8.

Figure 7-7: The menu bar consists of the File, Edit, Tools, and Help menus.

NOTE Selecting Tools ▸ Update NXT Firmware... launches a dialog that you can use to update your NXT's firmware. For more instructions on how to use this feature, consult pages 74 through 75 in your LEGO MINDSTORMS user guide.

the toolbar

The *toolbar* is essentially 12 clickable icons that represent various commands (Figure 7-8). You can select some of these commands in alternative ways, such as from the menu bar, but the toolbar provides a convenient way of accessing them. You're already familiar with the first two commands, New Program and Open Program. The third command, *Save Program*, saves your changes to the active program. With the exception of the Create My Block command (which is covered in Chapter 8), you'll learn how to use all of the other commands later in this chapter.

NOTE If you forget which command an icon represents, place your cursor over the icon and pause a moment. A tooltip will appear that specifies the underlying command.

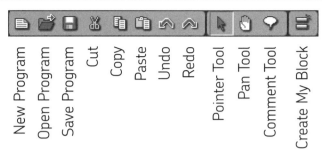

Figure 7-8: The toolbar consists of 12 clickable icons that represent various commands.

the work area

Consisting of a boundless grid of light gray squares, the *work area* is where you do all your programming (Figure 7-9). In other words, you drag the programming blocks into the work area.

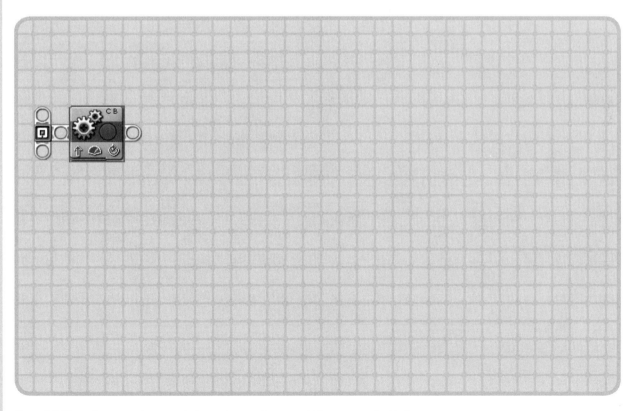

Figure 7-9: The real programming occurs in the work area, where you add the programming blocks.

the programming palette

The *programming palette* at the far left of the interface consists of three divisions—the Common palette, the Complete palette, and the Custom palette—which hold all the programming blocks (Figure 7-10). You can navigate among the palettes by clicking the tabs at the bottom, and you can add a programming block to a program by clicking the programming block on the palette and dragging that block's icon into the work area. The Common palette, which we'll cover in "The Common Palette" on page 79, contains the most frequently used programming blocks. The Complete palette contains all standard programming blocks, and the Custom palette contains programming blocks that you've made yourself or downloaded from the Internet. We'll discuss the Complete and Custom palettes in Chapter 8.

the configuration panel

Positioned in the lower-left corner of the screen, the *configuration panel* enables you to view and configure the options (if any) for the currently selected programming block. Figure 7-11 shows the configuration panel for the Move block with all the default options selected. Whenever you click a different programming block, the configuration panel changes to reflect that block's options. In addition, the configuration panel goes blank when you select more than one programming block (you can configure only one block at a time) or when no programming block is selected.

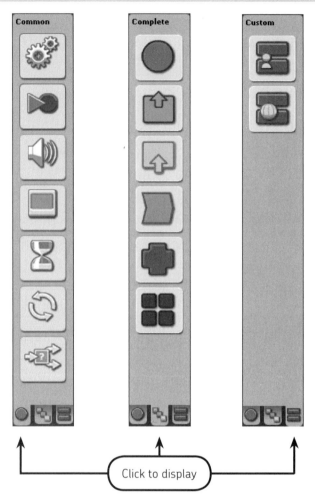

Figure 7-10: You can navigate among the Common palette, the Complete palette, and the Custom Palette by clicking the tabs at the bottom.

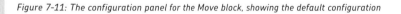

Figure 7-11: The configuration panel for the Move block, showing the default configuration

the controller

Positioned in the lower-right corner of the work area and shown in Figure 7-12, the *Controller* enables you to communicate with the NXT. It consists of five buttons, and Table 7-1 gives their names and functions.

Figure 7-12: The Controller consists of five buttons that you can use to communicate with the NXT.

table 7-1: the controller buttons

button	name	function
1	NXT Window button	Displays a dialog for managing NXT communications and memory
2	Download button	Downloads the active program to the NXT
3	Stop button	Stops any program running on the NXT
4	Download and Run button	Downloads the active program to the NXT and immediately executes it
5	Download and Run Selected button	Downloads *only* the currently selected programming blocks to the NXT and immediately executes them

NOTE The Download and Run Selected button is useful for testing portions of code within a program.

Clicking the NXT Window button displays the dialog in Figure 7-13, which has two tabs: Communications and Memory. The Communications tab (Figure 7-13, top) is selected by default and enables you to manage the connections between one or more NXTs and your computer. The Memory tab (Figure 7-13, bottom) enables you to manage the memory of the connected NXT. (For detailed information about this advanced feature, choose **Help ▸ Contents and Index...** to access the software's documentation, and then consult the "Files and Memory on the NXT" topic.)

The NXT Data panel is located to the right of the NXT Window and continuously provides information about the currently connected NXT, including its name, battery level, connection type, remaining memory, and firmware version. In this example, I've connected an NXT named *D's NXT* (short for *David's NXT*) through Bluetooth that has a battery level of 7.6 volts, 37.4KB of free storage, and firmware version 1.03. To rename an NXT, simply type a name in the text box next to the word *Name:* and then click the enter button to the right of that text box.

Figure 7-13: The NXT Window has a Communications tab for managing NXT-to-computer connections and a Memory tab for managing the currently connected NXT's memory.

NOTE You should occasionally defragment your NXT's memory to prevent potential errors. Clicking the Delete All button in the lower-left corner of the dialog in Figure 7-13 deletes all user-created programs on the NXT *and* defragments the NXT's memory. If you have important files on your NXT, make sure to back up to your computer first.

the little help window and the work area map

In the lower-right corner of the main screen is a small window with two tabs. The top tab (the question mark) is selected by default, displaying the Little Help Window (Figure 7-14). Whenever you place your cursor over an object in the work area, the Little Help Window displays information related to that object and links to more detailed information in the software's documentation. In Figure 7-14, I placed my cursor over a Move block, and the Little Help Window displayed information about using that particular block. Clicking the *More help »* link would take me to the section in the documentation about the Move block.

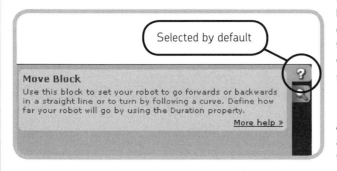

Figure 7-14: The Little Help Window gives you a little help by displaying descriptions of items in the work area and linking to more detailed documentation.

If you click the second tab (the magnifying glass), this same window will display the Work Area Map (Figure 7-15). Because NXT-G programs can easily extend beyond what you can see in the work area at one time, the Work Area Map enables you to view a thumbnail of the entire program. Clicking or clicking and dragging your cursor on the Work Area Map shifts your view of the work area.

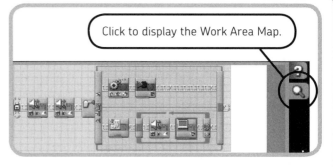

Figure 7-15: The Work Area Map enables you to view an entire program, regardless of its size.

fundamental NXT-G concepts

Before we examine the individual programming blocks, let's discuss some fundamental NXT-G concepts that are essential to the programming process: the starting point, the sequence beam, managing programming blocks in the work area, navigating the work area, and increasing program readability with comments.

the starting point

An NXT-G program begins with the programming block attached to the *starting point*, an object at the far left side of the work area (Figure 7-16). Notice that to the right of the starting point is the word *Start*, surrounded by four small marks that outline the shape of a block. Placing a block in this spot automatically attaches it to the starting point.

Figure 7-16: A program begins with the first block attached to the starting point.

sequence beams

An integral part of NXT-G is the *sequence beam*, which manages the *flow of control*, or the order of execution, for the programming blocks. The sequence beam stems from the starting point, and you must attach all of the programming blocks for a program to the sequence beam. The blocks will then execute one after another, in the order they appear on the sequence beam (Figure 7-17). For instance, the programming block on the far left attached to the starting point executes first, the next block on the sequence beam executes second, the block on the sequence beam after that executes third, and so on. This is known as *sequential execution*. It is possible to alter the flow of control by using certain programming blocks, however, and we'll discuss some of those later in this chapter.

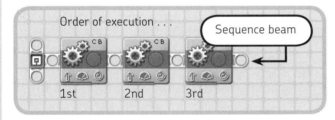

Figure 7-17: The sequence beam manages the flow of control through sequential execution.

NOTE Although you can place programming blocks in the work area without attaching them to a sequence beam, these blocks aren't considered to be part of the program, and the software ignores them when you download the program to the NXT.

In order to master NXT-G, one must master the use of the sequence beam. Therefore, we'll spend some time understanding exactly how the sequence beam functions and discussing some vital sequence beam techniques.

extending the sequence beam

In order to add another programming block to the sequence beam, the sequence beam must be extended. Fortunately, this is done automatically. For example, Figure 7-18 shows a sequence beam with one block attached, but what should you do if you want to add another one? Approaching the sequence beam with another block causes the sequence beam to extend automatically so that you can easily add the block to the program (Figure 7-19). The same holds true if you add a block between two other blocks.

Figure 7-18: This sequence beam has one programming block attached, and we'd like to add another one.

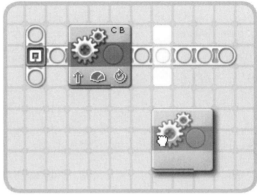

Figure 7-19: As we approach the sequence beam with another block, the sequence beam automatically extends to make room for the new block.

You can also adjust the sequence beam's length manually. If you position your cursor over a visible portion of the sequence beam, your cursor turns into a symbol with two arrows and lines. You can then click and drag the sequence beam to the right or left, resizing it as desired (Figure 7-20). In practice, though, you'll rarely use this technique.

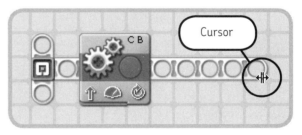

Figure 7-20: Extending the sequence beam manually

parallel sequence beams

You can also create *parallel sequence beams* that execute their own sets of programming blocks in addition to the blocks executing on the main sequence beam.[*] Hence, a program can simultaneously perform multiple tasks by using one or more parallel sequence beams. A parallel sequence beam can stem from the starting point or any other visible portion of a sequence beam. In fact, you can add a parallel sequence beam to a parallel sequence beam!

Consider that the starting point has three potential sequence beams protruding from it: one on the top, one on the side (the main sequence beam), and one on the bottom. If you position your cursor over the top or bottom one, you'll notice that your cursor turns into a reel. Click once to "attach" the sequence beam to your cursor. Next, use your cursor to draw out the sequence beam to the desired location and then double-click to finalize the position. Figure 7-21 shows a program with a main sequence beam and two parallel sequence beams stemming from the starting point. The main sequence beam and the lower parallel beam each have a Move block attached to them, and the upper parallel sequence beam is being drawn out.

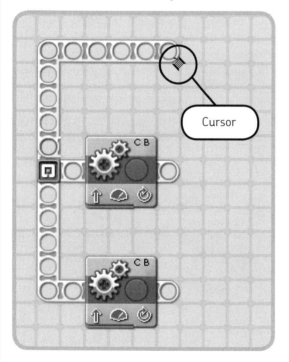

Figure 7-21: Creating a parallel sequence beam that stems from the starting point

To create a parallel sequence beam from anywhere other than the starting point, position your cursor over a visible portion of a sequence beam, press the SHIFT key to turn your cursor into the familiar reel, and then click once to attach the sequence beam to your cursor. Draw out the sequence beam with your cursor as before, double-clicking to finalize the position (Figure 7-22). Note that while creating the sequence beam, you can click once to pin it down where your cursor is positioned and then continue drawing it out. You can pin down a sequence beam in as many spots as you want. This technique also applies to drawing sequence beams out from the starting point, but you'll use it more often when creating sequence beams elsewhere in the program.

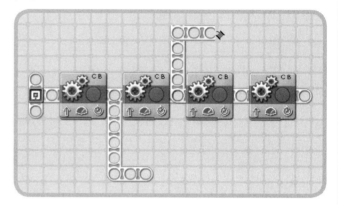

Figure 7-22: Creating parallel sequence beams within a program

automatic resizing of sequence beams

Finally, it's worth noting that anytime you remove or delete one or more programming blocks from a sequence beam, the sequence beam automatically reduces its size to accommodate the remaining blocks. This behavior is helpful because it prevents you from having to resize the sequence beam each time you change a program. But when there are gaps between programming blocks on a sequence beam, this behavior can sometimes lead to substantial changes. For example, in Figure 7-23 there is a gap of six round-holes between two Move blocks on the main sequence beam. If I remove the block at the tip of the sequence beam, the sequence beam automatically resizes to accommodate the remaining block. Keeping this characteristic of sequence beams in mind will help you better manage your programs.

---

[*] This technique is commonly known among programmers as *multithreading*.

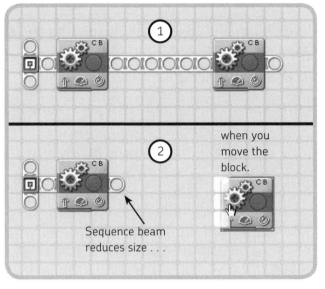

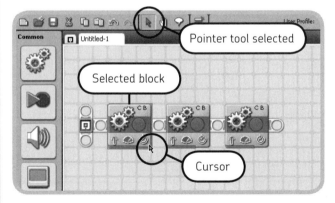

Figure 7-23: When you remove or delete a programming block from a sequence beam, the sequence beam automatically resizes.

managing programming blocks in the work area

As you know by now, an important part of NXT-G programming involves the management of programming blocks within the work area, such as moving blocks and deleting blocks. In order to manage blocks, you must first select them. And in order to select blocks, you must first select the *Pointer tool* from the toolbar by clicking the arrow icon. With the Pointer tool active, you can click a block to select it, which outlines the block in blue (Figure 7-24). To select

Figure 7-24: To select a programming block, select the Pointer tool from the toolbar and then click a block.

multiple programming blocks, use the Pointer tool to drag a selection rectangle over the blocks (Figure 7-25). You can also hold down the SHIFT key and click blocks one at a time to add them to the selection, but I usually find it easier and more efficient to use a selection rectangle.

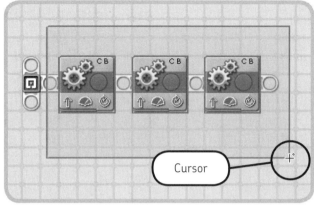

Figure 7-25: To select multiple programming blocks, drag the Pointer tool to create a selection rectangle.

Once you've selected one or more programming blocks, you can perform any of the following actions:

* You can *move* the selected block to a new location by clicking and dragging it.
* You can *delete* the selected block by clicking the cut icon on the toolbar, choosing Edit ▸ Cut, or pressing CTRL-X (Windows) or CMD-X (Mac).
* You can *copy* the selected block by clicking the copy icon on the toolbar, choosing Edit ▸ Copy, or pressing CTRL-C (Windows) or CMD-C (Mac).
* You can then *paste* the copied block by clicking the paste icon on the toolbar, choosing Edit ▸ Paste, or pressing CTRL-V (Windows) or CMD-V (Mac).

NOTE There is also a keyboard shortcut for copying *and* pasting. Hold down CTRL (Windows) or OPTION (Mac) and then drag a selected block to produce a duplicate.

If you make a mistake, you can *undo* actions by clicking the undo icon on the toolbar, choosing Edit ▸ Undo, or pressing CTRL-Z (Windows) or CMD-Z (Mac). You can also *redo* undone actions by clicking the redo icon on the toolbar, choosing Edit ▸ Redo, or pressing CTRL-SHIFT-Z (Windows) or SHIFT-CMD-Z (Mac).

navigating the work area

While working on a program, you'll often need to adjust your view of the work area to see all of the various programming blocks. The Work Area Map discussed earlier in "The Little Help Window and the Work Area Map" on page 74 works well for navigating large programs, but it's easier to use the *Pan tool* (the hand icon on the toolbar) when working with small programs or with portions of code in large programs. Once you've selected the Pan tool, your cursor changes into a hand when hovering over the work area, and you can click and drag in the work area to shift your view (Figure 7-26).

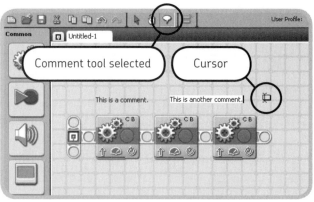

Figure 7-27: Select the Comment tool from the toolbar to add a comment to a program.

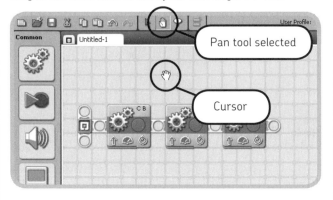

Figure 7-26: Selecting the Pan tool from the toolbar enables you to click and drag in the work area to shift your view.

NOTE When using the Pointer tool, you can temporarily activate the Pan tool by holding down CTRL-SHIFT (Windows) or OPTION-SHIFT (Mac).

increasing program readability with comments

Adding comments to your programs increases their readability by helping others (and your future self) understand how they function. Documenting NXT-G programs with comments is particularly important given their graphical nature; the purpose of a programming block or a group of programming blocks within a program isn't always obvious, and comments can help you make sense of it.

To add a comment in NXT-G, select the *Comment tool* (the word balloon icon on the toolbar) and then click anywhere in the work area. A small white box appears in which you can type your comment. Figure 7-27 shows one comment placed above the first programming block and a second comment being typed to its right.

To select and move a comment within the work area, the Pointer tool must be active. In addition, another way to

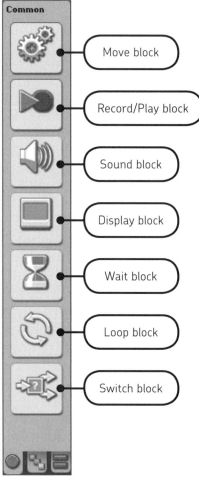

Figure 7-28: The Common palette offers the seven most frequently used programming blocks, which are collectively called the Common blocks.

NXT-G KEYBOARD AND MOUSE SHORTCUTS

Table 7-2 presents keyboard and mouse shortcuts for the most common NXT-G operations. With the exception of one shortcut that you can use to switch between open programs and the main screen, all of these shortcuts are discussed in this chapter.

table 7-2: NXT-G keyboard and mouse shortcuts

operation	shortcut (windows)	shortcut (mac)
New program	CTRL-N	CMD-N
Open program	CTRL-O	CMD-O
Cut	CTRL-X	CMD-X
Copy	CTRL-C	CMD-C
Paste	CTRL-V	CMD-V
Undo	CTRL-Z	CMD-Z
Redo	SHIFT-CTRL-Z	SHIFT-CMD-Z
Switch between open programs and main screen	CTRL-TAB	CMD-TAB
Temporary Pan tool	Hold down CTRL-SHIFT	Hold down OPTION-SHIFT
Add comment	Double-click in work area	Double-click in work area
Select multiple blocks (one at a time)	Hold down the SHIFT key and click a block	Hold down the SHIFT key and click a block
Duplicate selected block	Hold down the CTRL key, place cursor over a selected block, and drag	Hold down the OPTION key, place cursor over a selected block, and drag

add comments to your program is by selecting the Pointer tool and then double-clicking within the work area, which likewise launches a white comment box. I've found that this is usually a more efficient means of adding comments rather than by using the Comment tool.

the common palette

The *Common palette* contains the most frequently used programming blocks, which are collectively known as the *Common blocks*. These include the Move block, the Record/Play block, the Sound block, the Display block, the Wait block, the Loop block, and the Switch block (Figure 7-28). You can create a large number of functional programs using only the Common blocks.

the move block

One of the most commonly used programming blocks of all is the *Move block*, which controls the servo motors. The Move block is actually a specialized block for driving simple mobile robots like TriBot or Zippy-Bot (see Chapter 11), but you can configure the Move block to control motors in any type of robot. Figure 7-29 points out four symbols on the Move block that tell us at a glance how it's basically configured; these symbols change depending on the block's configuration. Table 7-3 describes each of the symbols.

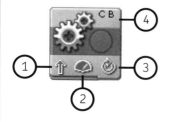

Figure 7-29: The Move block

table 7-3: the move block's symbols

symbol	purpose	default
1	Shows direction of travel or direction of motor rotation	Forward
2	Shows power level	75 percent
3	Shows unit of measure for duration	Rotations
4	Shows selected output ports	Output ports C and B

Figure 7-30 displays the Move block's default configuration panel. There are six different parameters to configure: Port, Direction, Steering, Power, Duration, and Next Action. On the far left side of the panel are *feedback boxes* that help you to measure motor movements. In the following sections, we'll examine each of these Move block parameters and the feedback boxes.

the port parameter

Check boxes A, B, and C correspond to output ports A, B, and C on the NXT. You indicate which output ports (motors) the Move block should control by checking these boxes. You can select one, two, or three ports; how many you select depends on the type of robot you're programming. Let's consider three general rules.

First, you should control two output ports with a Move block only when driving a mobile robot like TriBot that uses two motors to move, one on each side of the robot. With two ports checked, the Move block automatically synchronizes the motors so that the robot can drive in a straight line, and it enables a "steering" feature that we'll observe in a moment.

NOTE If you configure a Move block to control two ports but only use one of them (i.e., connect a motor to only one of the ports), the motor will not perform correctly—or at all. Why? The Move block is continuously attempting to synchronize the movement of two motors, one of which *isn't* moving because it's not there. Therefore, the connected motor cannot move either.

Second, you should control three output ports with a Move block only when driving the previously described type of mobile robot and operating an "extra" motor at the same time (i.e., a motor that serves some purpose other than driving). When three ports have been checked, the Move block assumes that output ports B and C are driving the robot and automatically synchronizes them; since you cannot change this default, ensure that you've attached the driving motors to output ports B and C on the NXT.

Figure 7-30: The Move block's configuration panel (default configuration)

Third, you should control one output port with a Move block whenever you *aren't* driving the previously described type of mobile robot so that the synchronization and steering features aren't enabled. A stationary robot like a robotic arm or a mobile robot like a tricycle cannot use those features because of differences in its design. Instead of using a Move block to control both ports B and C, for example, you might have a Move block that controls port B, followed by another Move block that controls port C.

the direction parameter

The *Direction parameter* controls the direction of rotation for the designated motor(s): forward, reverse, or stop. Selecting the first option (the up arrow) causes the motor shafts to spin in the forward direction, and selecting the second option (the down arrow) causes the motor shafts to revolve in the reverse direction. Figure 7-31 demonstrates the forward and reverse directions of the servo motors. Selecting the third option stops all the motors designated in the Port parameter.

NOTE Importantly, the forward option doesn't always drive a mobile robot forward. Depending on a robot's design, the forward option may drive the robot backward; you'll see an example of this in Chapter 11.

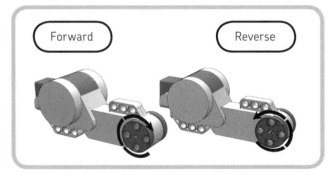

Figure 7-31: The forward and reverse directions of the servo motors

the steering parameter

Enabled only when *two* output ports are selected, the *Steering parameter* allows you to steer the robot in any direction. When the slider is positioned exactly in the middle, as shown in Figure 7-32, the Move block synchronizes the motors so that the robot moves in a straight line. If you move the slider left or right, the robot "steers" in that direction by changing the speeds of the motors or even reversing a motor's direction of rotation. The Move block assumes by default that the motor on port C is on the left side of the robot and that the motor on port B is on the right side of the robot. If this isn't the case, use the two drop-down menus to specify which motors are on which sides of the robot. In addition, if you select the reverse option in the Direction parameter, the Steering parameter changes to let you steer in reverse.

Figure 7-32: The Steering parameter on the Move block's configuration panel

the power parameter

Use the *Power parameter* to specify the power level for the motors, which ranges from 0 percent to 100 percent. You can either type the power level in the text box or move the slider with your mouse.

the duration parameter

Use the *Duration parameter* to specify the duration for which the motors should run in seconds, degrees, or rotations; you can also choose *unlimited*. When measuring duration in degrees, 360 degrees is one complete rotation of the motor's shaft; when measuring in rotations, one rotation is one complete rotation of the motor's shaft.

the next action parameter

Use the *Next Action parameter* to specify what the motors should do when the Move block has finished its action. Choosing *Brake* (the default) applies power to prevent the motors' output shafts from turning; choosing *Coast* simply cuts off the power supply going to the motors.

feedback boxes

A handy feature of the Move block is the *feedback boxes* that measure and display the movements of motors in degrees (Figure 7-33). The NXT must be connected to a computer to enable the feedback boxes. Whenever you manually move the motor shafts or run a program that powers the motors, the corresponding feedback boxes update their displays. The feedback boxes begin by displaying 0 for each output port. From then on, forward motion adds to the reported number and reverse motion subtracts from the reported total. Clicking the small *R* button resets the feedback box values to 0.

Figure 7-33: The feedback boxes measure the movements of the motors and display them in degrees.

The usefulness of the feedback boxes lies in their ability to help you determine the Duration parameter. For example, let's say that you want a robot to move from point A to point B, but you don't know how far to tell the robot to travel. You can determine this amount by pushing the robot from point A to point B by hand and then checking the feedback boxes for the answer in degrees.

the record/play block

The *Record/Play block* has, as its name suggests, two functions: recording the movements of motors and playing recorded movements of motors. The Record/Play block displays only one symbol at a time, indicating whether the robot is recording an action or playing an action (the symbol for Record mode is shown in Figure 7-34).

Figure 7-34: The Record/Play block

Figure 7-35 shows the configuration panel for the Record/Play block while in Record mode. There are four parameters to configure:

Action Specify whether you are recording or playing motor movements by selecting either the Record radio button (selected by default) or the Play radio button.

Name Type a name for the file that will store the motor movements. The default name is *RobotAction*.

Recording Specify the output ports from which to record motor movements: ports A, B, or C. You can check as many or as few as you want. Ports B and C are checked by default.

Time Specify (in seconds) how long to record the motor movements. The default value is 30 seconds, but you can adjust the settings from 0 to 2,147,483 by either typing directly in the text box or clicking the up and down arrows to the right of the text box.

Figure 7-35: The Record/Play block configuration panel in Record mode

If you select Play in the Action parameter, the configuration panel changes as shown in Figure 7-36. There are two different parameters to configure in Play mode:

Name You can type the name of a movement file to play in this text box, or if you select one from the File parameter, it appears here.

File Here you can select from a list of movement files already on the NXT. The NXT must be connected to your computer to enable this parameter.

Figure 7-36: The Record/Play block configuration panel in Play mode

the sound block

The *Sound block* can play either a sound file or a tone on the NXT's speaker. As Figure 7-37 shows, the Sound block has three symbols on it. Table 7-4 describes each of these symbols.

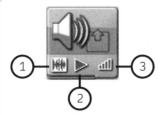

Figure 7-37: The Sound block

table 7-4: the sound block's symbols

symbol	purpose	default
1	Shows if a sound file or tone will play	Sound file
2	Shows if block starts or stops a sound	Start
3	Shows the volume	75 percent

Figure 7-38 shows the configuration panel for a Sound block configured to play a sound file. There are six parameters to configure:

Action   Specify whether the block plays a sound file or a tone.

Control   Specify whether the block plays a sound or stops sound.

Volume   Specify the volume as a percentage of the maximum volume: 0 percent, 25 percent, 50 percent, 75 percent, or 100 percent. You can either drag the volume slider with your cursor or type a percentage in the text box to the right of the slider.

Function   Specify if the block should continuously repeat its sound. If you leave the Repeat check box unchecked, the sound will only play once. If you check this box, the sound will continuously repeat, and the Wait parameter will be disabled.

File   Specify which sound file to play by selecting one from the list. Clicking a sound file plays the sound on your computer so that you can hear it first.

Wait   Specify if the program should wait for the Sound block to finish playing its sound before allowing the next block on the sequence beam to execute. If this box is unchecked (or disabled), the next block on the sequence beam can immediately execute, even if the Sound block isn't finished. However, the Sound block will immediately stop if *another* Sound block wants to play a sound.

NOTE   Sound files are very large, at least in terms of the NXT's available memory. For this reason, I recommend that you don't use more than two or three different sound files in a program. As an alternative, you can play tones, which don't require nearly as much memory.

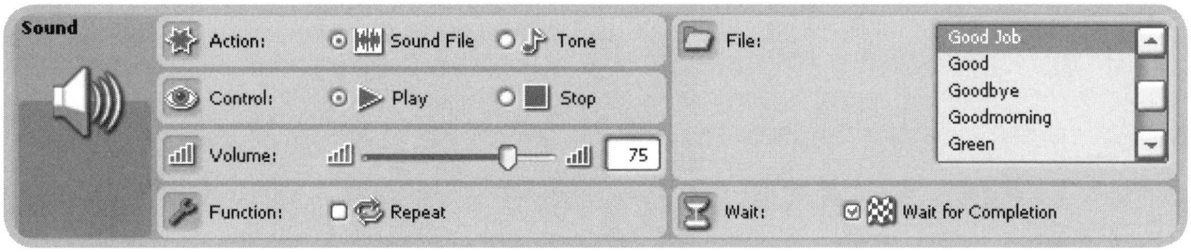

Figure 7-38: The Sound block's configuration panel for playing a sound file

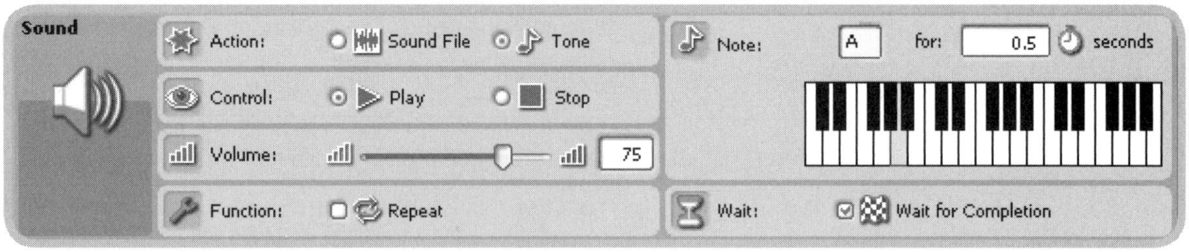

Figure 7-39: The Sound block's configuration panel for playing a tone

Figure 7-39 shows the configuration panel for a Sound block configured to play a tone. The only difference is that there is now a Note parameter in the upper-right corner instead of a File parameter. Using the *Note parameter*, you can specify a tone by pressing a key on the keyboard. The tone that corresponds to the key you pressed will sound on your computer, and its note (A, B, C, etc.) is shown above the keyboard. Directly above the keyboard is also a text box for specifying the length of time for which the tone should play.

In order to create music with Sound blocks that play tones, you need to position several Sound blocks in a row as shown in Figure 7-40.

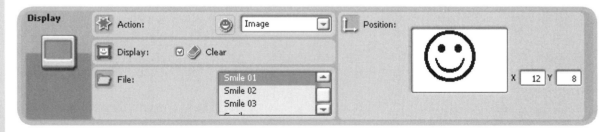

Figure 7-40: Three Sound blocks playing tones one after the other

the display block

The *Display block* can display an image, text, or drawing on the NXT's LCD. The Display block has one symbol on it that indicates whether it's configured to display an image, text, or a drawing (the image symbol is shown in Figure 7-41).

Figure 7-41:
The Display block

Figure 7-42 shows the configuration panel for the Display block configured to display an image, which is the default. There are four parameters to configure:

Action Specify whether the block should display an image, text, or a drawing, or if it should reset the display. When the display is *reset*, the LCD displays the regular menu.

Display Specify whether the block should clear the display of previous images, text, and drawings.

File Specify which image to display on the NXT's LCD.

Position Specify the position of the image on the NXT's LCD by either clicking the preview screen or typing the x- and y-coordinates in the text boxes to the right of the preview screen.

Figure 7-42: The Display block's configuration panel for displaying an image

Figure 7-43 shows the configuration panel for the Display block when set to display text. There are two parameters to configure that are different or configured differently:

Text   Specify what text you want to display on the NXT's LCD by typing it in the text box.

Position   Specify where the Display block should display the text on the NXT's LCD. To do this, you can click a position on the preview screen, type the x- and y-coordinates into the text boxes, or select a line (1–8) on which to display the text.

Figure 7-44 shows the configuration panel for the Display block when set to display a drawing. There are two parameters to configure that are different or configured differently:

Type   Use the drop-down menu to indicate whether you would like to display a single point (shown in Figure 7-44), a line, or a circle.

Position   Specify where the Display block should display the point, line, or circle on the NXT's LCD. Depending on which you choose, there will be different options.

the wait block

The *Wait block* tests for a specified condition, such as a pressed touch sensor, and doesn't allow the program or sequence beam to continue until that condition has been satisfied. Figure 7-45 shows a Wait block configured to wait for one second (i.e., to test to see if one second has passed) before allowing the next block on the sequence beam to execute.

Figure 7-45: The Wait block

There are nine main ways to configure the Wait block. As shown in Figure 7-46, positioning your cursor over the Wait block on the Common palette reveals a sub-palette from which you can select the five most common configurations: wait for a specified amount of time, wait for a specified touch sensor reading, wait for a specified light sensor reading, wait for a specified sound sensor reading, and wait for a specified ultrasonic sensor reading.

Figure 7-46: The Wait block's sub-palette gives you access to the five most common configurations.

Display   Action: [T] Text ▼   Position:

Display: ☑ Clear

Text: Mindstorms NXT   Mindstorms NXT X [12] Y [8]

Line: [7 ▼]

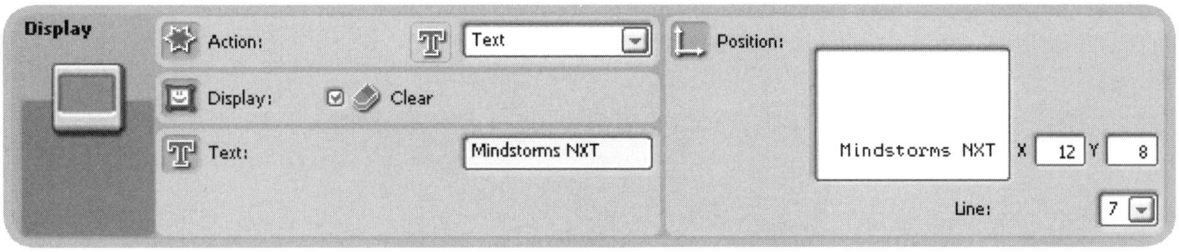

Figure 7-43: The Display block's configuration panel for displaying text

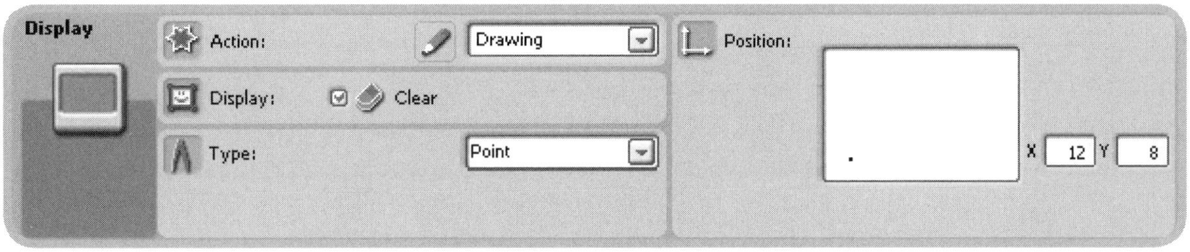

Display   Action: [✎] Drawing ▼   Position:

Display: ☑ Clear

Type: Point ▼   .   X [12] Y [8]

Figure 7-44: The Display block's configuration panel for displaying a drawing

Since the NXT-G documentation exhaustively documents each of the nine possible configurations, we'll look at just one: the touch sensor configuration. Figure 7-47 shows a simple program that starts off with a Move block that drives the robot for the Unlimited duration. The following Wait block keeps the robot driving until a touch sensor on input port 1 of the NXT is pressed. Once the touch sensor has been pressed, another Move block stops the motors. For this wait block configuration, there are four parameters to configure: Control, Sensor, Port, and Actions.

Control   Specify whether the Wait block should wait for a time limit or a sensor reading to be satisfied.

Sensor   You can select from eight different options: Touch Sensor, Light Sensor, Sound Sensor, Ultrasonic Sensor, NXT Buttons, Receive Message, Timer, or Rotation Sensor. The last "sensor" listed is the built-in rotation sensor in the servo motors, and the three preceding "sensors" are built in to the NXT.

Port   Specify which input port the touch sensor is using on the NXT.

Action   Specify which condition to test for: the touch sensor is pressed, the touch sensor is released, or the touch sensor is bumped (i.e., pressed and then released).

NOTE Depending on which of the nine Wait block configurations you choose, different symbols will appear on the Wait block and different parameters will need to be configured. Consult the NXT-G documentation for more information.

the loop block

The *Loop block*, which belongs to a category of programming blocks called *Flow blocks*, repeats a series of blocks until a specified condition is satisfied, such as a pressed touch sensor. This is one of the blocks that can alter the flow of control (bringing about a *transfer of control*) by changing the order of execution for the blocks on a sequence beam. Place the blocks you want repeated onto the sequence beam inside of the Loop block. In Figure 7-48, the Loop block is set to its default configuration in which it repeats blocks forever, as signified by the infinity symbol on the block.

Figure 7-48: The Loop block

Using the Control parameter on a Loop block's configuration panel, you can configure the Loop block to test for five main types of conditions: Forever, Sensor, Time, Count, and Logic. The *Forever option* repeats blocks forever. The *Sensor option* repeats blocks until a sensor condition is met. The *Time option* repeats blocks until a certain amount of time has passed. The *Count option* repeats the blocks a specified number of times. The *Logic option* involves the use of a *data wire* (logic and data wires are both covered in Chapter 8).

Let's look at an example. Figure 7-49 shows a Loop block configured to repeat three blocks (Move, Record/Play,

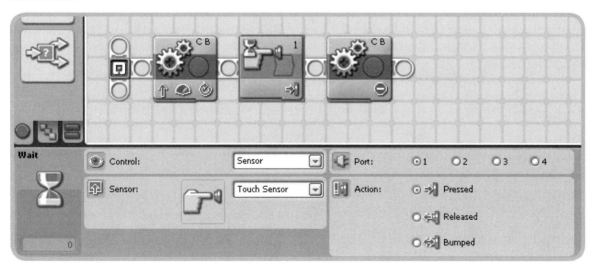

Figure 7-47: As shown in its configuration panel, the Wait block in this program does not allow the program to continue until the touch sensor connected to input port 1 is pressed.

and Sound) until a touch sensor is pressed. In this case, there are five parameters to configure:

Control   Specify the type of condition the Loop block will test for: Forever, Sensor, Time, Count, or Logic.

Sensor   You can select from eight different options: Touch Sensor, Light Sensor, Sound Sensor, Ultrasonic Sensor, NXT Buttons, Receive Message, Timer, or Rotation Sensor. (These same options are available for the Wait block.)

Show   Specify whether to include a plug for a data wire. (Data wires are covered in detail in Chapter 8.)

Port   Specify which input port the touch sensor is using on the NXT.

Action   Specify which condition to test for: the touch sensor is pressed, the touch sensor is released, or the touch sensor is bumped (i.e., pressed and then released).

NOTE There is a feedback box in the left portion of the Loop block's configuration panel that can display sensor readings when the NXT is connected to a computer (Figure 7-49). Both the Wait block and the Switch block can also display sensor readings in a feedback box.

the switch block

The *Switch block*, another Flow block, is a highly important programming block that enables your program to choose between two alternative "paths" of code based on a specified condition.[*] In Figure 7-50, the Switch block is set to its default configuration, in which it bases its decision on the

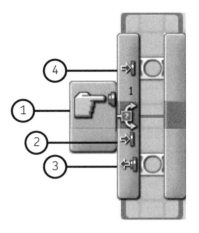

Figure 7-50: The Switch block

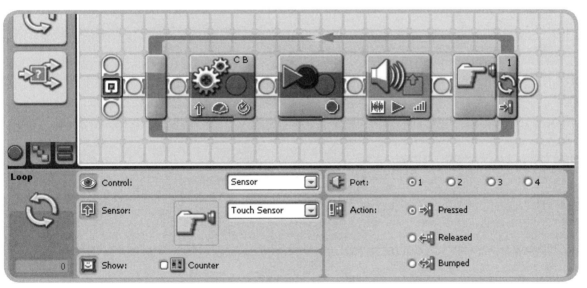

Figure 7-49: As shown in its configuration panel, this Loop block repeats the three blocks until a touch sensor connected to input port 1 is pressed.

---

[*] If you're familiar with text-based programming languages, you'll recognize the Switch block as a type of if-then structure.

reading of a touch sensor. Table 7-5 describes the symbols on this Switch block.

table 7-5: the switch block's symbols

symbol	purpose	default
1	Shows type of sensor	Touch sensor
2	Shows specified condition	Touch sensor pressed
3	Shows the condition that causes the blocks on the bottom to execute	Touch sensor released
4	Shows condition that causes the blocks on the top to execute	Touch sensor pressed

Using the Control parameter on a Switch block's configuration panel, you can configure the Switch block to test for two main types of conditions: Value or Sensor. The *Value* condition offers three types of values: Logic, Number, and Text. All of these require the use of data wires, which we'll

discuss in Chapter 8. The *Sensor* condition provides the same options and parameters as in the Wait and Loop blocks.

NOTE When configured to test for a Value that is either Number or Text, the Switch block can have more than two possible paths of code. The flat view option must be turned off, however (see the Display parameter).

Let's look at an example that utilizes the Switch block. Figure 7-51 shows the same Switch block in Figure 7-50 but with blocks inside of it. If the touch sensor is pressed, any blocks on the top sequence beam execute; if the touch sensor is not pressed, any blocks on the bottom sequence beam execute. Figure 7-51 also shows the Switch block's configuration panel. For this configuration, there are five parameters to configure:

Control Specify whether the block should test for a sensor or a value.

Sensor You can select from eight different options: Touch Sensor, Light Sensor, Sound Sensor, Ultrasonic Sensor, Rotation Sensor, NXT Buttons, Timer, or Receive Message. (These same options are available for the Wait and Loop blocks.)

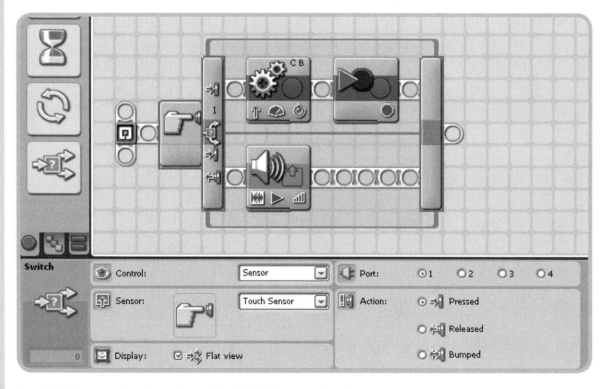

Figure 7-51: As shown in its configuration panel, the Switch block executes the code on its top sequence beam if the touch sensor connected to input port 1 is pressed and the code on its bottom sequence beam if the same touch sensor is released.

Display Specify whether the Switch block should be on flat view or not. In flat view (shown), both sequence beam paths are visible. With flat view turned off, the Switch block presents tabs that you must click to view the different sequence beam paths.

Port Specify which input port the touch sensor is using on the NXT.

Action Specify which condition to test for: the touch sensor is pressed, the touch sensor is released, or the touch sensor is bumped (i.e., pressed and then released).

NOTE For more information about the Switch block and the Loop block and all of their possible configurations, consult the NXT-G documentation included on the MIND-STORMS NXT software.

conclusion

Programming an NXT robot is quite different from building an NXT robot, but both activities are incredibly fun, and they allow you to unleash your creativity in amazing ways. In this chapter we explored the official NXT-G programming language, a powerful but straightforward graphical programming language. First, we discussed how to manage user profiles and start NXT-G programs. Next, we thoroughly examined the NXT-G interface and considered some fundamental NXT-G programming concepts. Finally, you learned how to use the programming blocks in the Common palette—the Common blocks. In the next chapter, we'll discuss advanced NXT-G concepts and you'll learn how to use the Complete palette and the Custom palette.

advanced NXT-G programming

An important step toward mastering the NXT set is becoming proficient in one or more programming languages for the NXT. With basic programming skills, you can successfully program NXT robots to perform basic tasks like grabbing a ball. But programming NXT robots to perform more complex tasks—such as finding a ball, grabbing it, and then transporting it to a designated location—requires more advanced programming skills.

Therefore, in this chapter we'll continue our discussion of NXT-G and examine its advanced features. We'll begin with data wires, which connect programming blocks and transmit information between them. Next, we'll learn how to use the programming blocks in the Complete palette, and then we'll explore the Custom palette and creating our own programming blocks. Finally, we'll briefly discuss how to expand the potential of NXT-G by importing new types of blocks with Dynamic Block Update, an official add-on.

data wires

Our first task is learning how to use *data wires* to transmit data between programming blocks (Figure 8-1). With the exception of the Wait block, all programming blocks can use data wires to one degree or another, and a number of blocks *require* the use of data wires. I recommend that you read this section carefully; working with data wires is more difficult than working with programming blocks. If you attempt to use data wires without an understanding of the basic concepts surrounding them, you can easily run into problems.

First and foremost, the foundation of data wires is the *data hub*, which you can access by clicking the small tab in the lower-left corner of a block (Figure 8-2). Some blocks immediately show all or part of their data hub when placed in the work area, while other blocks' data hubs are closed by default.

Along the length of each data hub is one or more *data plugs*, which come in two main types: *input plugs* and *output plugs* (Figure 8-3). Input plugs are always on the left side of a data hub, and they *receive* data from other blocks through data wires. Output plugs are always on the right side of a data hub, and they *send* data to other blocks through data wires. From each block's perspective, a data wire connected to one of its input plugs is an *input data wire*, and a data wire connected to one of its

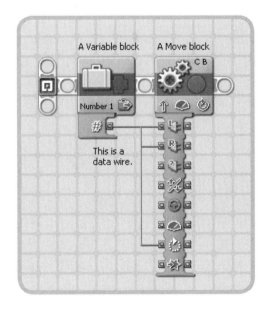

Figure 8-1: You can use data wires to transmit information between programming blocks.

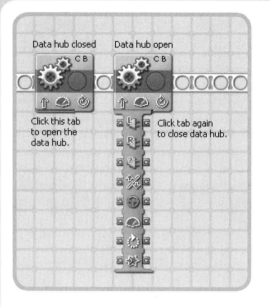

Figure 8-2: A block's data hub can be accessed by clicking
the small tab in the lower-left corner of the block.

output plugs is an *output data wire*. Not all blocks have both input and output plugs on their data hubs, however, and the type and number of plugs may depend on the block's configuration.

NOTE The Wait, Loop, and Switch blocks don't have data hubs, but the Loop block and Switch block do have data plugs that appear for certain configurations.

If you position your cursor over a data plug, the cursor turns into a reel. Clicking the plug once attaches a data wire to your cursor, and moving the cursor away from the plug "unrolls" the data wire. Clicking a data plug on another block completes the data wire connection (Figure 8-4). Even though creating a data wire may be easy, creating a meaningful and valid data wire requires an understanding of four concepts: data plug characteristics, the wire path, transmitting the data types, and broken data wires.

NOTE You can remove a data wire by clicking the connecting input plug. In some cases, however, you may need to click the data wire to select it, and then cut or delete it as you would a block.

data plug characteristics

A data plug has three characteristics that determine how you use it. First, a data plug relates to a particular aspect of its block, symbolized by an icon on the block's data hub. For example, the uppermost pair of input and output plugs on the Move block is collectively called the *Left Motor plug* and is symbolized by a plug with the letter *L* (visible in Figure 8-4).

Second, a data plug uses a specific data type: number, logic, or text. *Number data* simply consists of numbers such as 3, 12.8, or –150. *Logic data* involves only two values: true and false. *Text data* consists of characters such as letters, punctuation marks, and even numbers. Examples of text data include *robot*, *123bleep*, and *5!6gv*. We can refer to a plug as a *number plug*, *logic plug*, or *text plug* if it uses

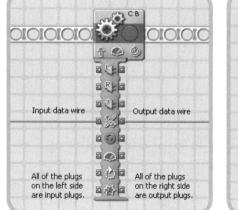

Figure 8-3: Input plugs, output plugs, input data wires, and output data wires on a block's data hub

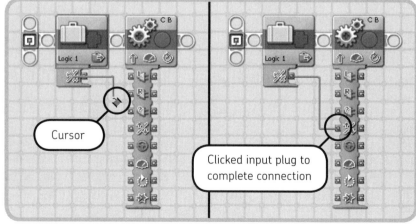

Figure 8-4: Clicking a data plug and then clicking another data plug on a different block creates a data wire connection.

number data, logic data, or text data, respectively. A little later in this chapter, we'll see how to properly transmit these different data types.

Third, a data plug often accepts only a certain range of values. For example, the Left Motor plug accepts number data in the range of 1 to 3. If the number data supplied is a 1, the left motor will be on output port A; if the number data supplied is a 2, the left motor will be on output port B; and if the number data supplied is a 3, the left motor will be on output port C. When an input plug receives data that's outside its accepted input range, it ignores the data if its range consists of only a few numbers (e.g., 1–3); and if its input range consists of a greater number of values (e.g., 1–100), it actually changes the data to fit within the range.

To determine these three pieces of information about a data plug, you'll need to consult the block's *data hub chart*. Figure 8-5 shows the data hub chart for the Move block. Of particular significance is the column beneath the words *What the Values Mean*. This information tells you how a data plug "reads" a particular value. You can find each block's data hub chart at the very bottom of the block's documentation on the software. Remember that if you place your cursor over a block in the work area, the Little Help Window displays a link to documentation for that block.

This chart shows the different characteristics of the plugs on the Move block's data hub:

	Plug	Data Type	Possible Range	What the Values Mean	This Plug is Ignored When...
	Left Motor	Number	1 - 3	1 = A, 2 = B, 3 = C	
	Right Motor	Number	1 - 3	1 = A, 2 = B, 3 = C	
	Other Motor	Number	1 - 3	1 = A, 2 = B, 3 = C	
	Direction	Logic	True/False	True = Forwards, False = Backwards	
	Steering	Number	-100 - 100	< 0 = Steer towards left motor, > 0 = Steer towards right motor	
	Power	Number	0 - 100		
	Duration	Number	0 - 2147483647	Depends on Duration Type: Degrees/Rotations = Degrees, Seconds = Seconds	Duration Type = Unlimited
	Next Action	Logic	True/False	True = Brake, False = Coast	Duration Type = Unlimited, Steering not equal to zero (this may only be temporary, pending firmware fix for this not to be ignored)

Figure 8-5: The data hub chart for the Move block

the wire path

When a piece of data initially leaves an output plug and travels over a data wire to one or more blocks, it's following a *wire path*. A wire path may consist of a single data wire or multiple data wires. In order to create functional wire paths, you should follow two important guidelines.

First, every wire path must connect to at least one output plug and one input plug. You cannot connect data wires only to input plugs or only to output plugs. Moreover, a wire path must always begin at an output plug *without* a corresponding input plug. A "lone" output plug like this sends data from its block, and, of course, a wire path must begin at the data source. An example of one of these output plugs is on the Variable block (which we'll discuss later) in Figure 8-6.

Second, you can extend a wire path to reach more than one block by using corresponding pairs of input and output plugs. In Figure 8-6, the first Move block receives data from the Variable block, this input data passes *unchanged* from the Move block's input plug to its corresponding output plug, and then a data wire connected to that output plug sends the data to the next block. You *cannot* use an output plug if it has a corresponding input plug without a data wire connected to it; the output plug serves only to transmit input data to other blocks.

NOTE If there are unused plugs after you've made your data wire connections, you can click a block's data hub tab so that it resizes its data hub to show only the currently used data plugs. Opening or closing data hubs often causes the software to rewire or reposition sections of the wire path, but the resulting wire path still functions in the same way.

In summary, the wire path in Figure 8-6 consists of two data wires and transmits data to two blocks. For instance, if the Variable block sent the value 3 (number data), both Move blocks would receive the value 3. Theoretically, we could extend the wire path to transmit data to an unlimited number of blocks.

transmitting the data types

A data wire must connect plugs of the same data type. For example, a data wire connected to a number plug on one block can connect only to number plugs on another block. You can verify a plug's data type by consulting the block's data hub chart. Fortunately, however, the data wires are color-coded. Data wires that carry number data are yellow, data wires that carry logic data are green, data wires that carry text data are orange, and "broken" data wires are gray (Figure 8-7). We discuss broken data wires next.

broken data wires

A data wire "breaks" when you've made an invalid connection between two data plugs, and you cannot download a program to your NXT if that program contains broken wires. Broken data wires generally result from three types of errors: data type mismatch, missing input, or too many inputs. We'll briefly observe each of these errors. Note that if you create a broken data wire, you should simply undo the action instead of attempting to remove the wire by deleting it—you might make the problem worse!

NOTE If you position your cursor over a broken data wire, the Little Help Window will tell you which type of error caused the data wire to break.

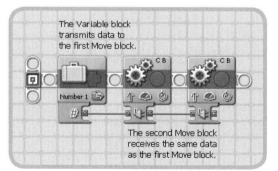

Figure 8-6: This wire path begins with a Variable block and extends to the second Move block.

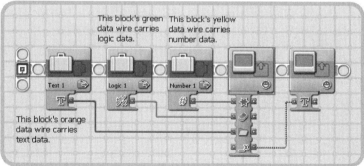

Figure 8-7: Data wires are yellow when transmitting number data, green when transmitting logic data, and orange when transmitting text data. Broken data wires are gray and result from mismatching data types.

data type mismatch

Mismatching data types is one of the most common mistakes. We already discussed this topic in "Transmitting the Data Types" on the previous page, but it's worth repeating: A data wire *must* connect plugs of the same data type.

missing input

Another common mistake is attaching a data wire to an output plug when that output plug's corresponding input plug has no data source (Figure 8-8). I emphasized this topic in "The Wire Path" on the previous page. Remember that with a pair of corresponding input and output plugs, the output plug can only send the data received by the input plug. If you want to begin a wire path, you use an output plug *without* a corresponding input plug.

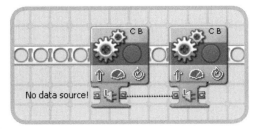

Figure 8-8: This broken data wire is missing an input.

too many inputs

When an input plug receives data from more than one output plug, the wire path has too many inputs (Figure 8-9). A wire path can connect to multiple input plugs, but it must begin at one output plug.

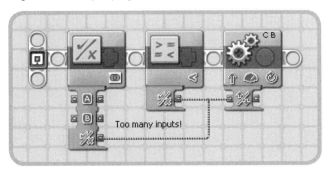

Figure 8-9: The Move block on the far right has too many inputs.

NOTE The NXT-G documentation on the software discusses a fourth type of error, a *cycle*, which occurs when a wire path "visits the same block twice." However, the software prevents you from creating cycles by not allowing certain data wire connections.

the complete palette

We're now prepared to discuss the *Complete palette*, the second of the three programming palette divisions, which contains every type of standard programming block—30 in all. The Complete palette groups its programming blocks into six categories: Common blocks, Action blocks, Sensor blocks, Flow blocks, Data blocks, and Advanced blocks (Figure 8-10).

NOTE Appendix B summarizes the basic information about each of the 30 standard programming blocks.

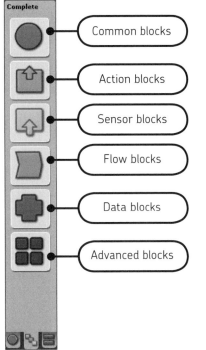

Figure 8-10: The Complete palette includes six categories of programming blocks.

the common blocks

The *Common blocks* in the Complete palette are the same as the Common blocks in the Common palette, which was discussed in Chapter 7. If you plan to use solely the Common blocks for a program, you should use the Common palette because you can more conveniently access them there.

the action blocks

As you might have guessed, the *Action blocks* instruct your robot to carry out a particular action. Positioning your cursor over the Action blocks category on the Complete palette displays a sub-palette with four Action blocks: the Motor block, Sound block, Display block, and Send Message block (Figure 8-11). As we discuss how to configure the Action blocks through their configuration panels, remember that you can control those same options dynamically with data wires. You would need to open their data hubs, however, because they're closed by default.

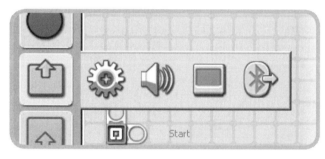

Figure 8-11: The Action block's sub-palette includes the Motor block, the Sound block, the Display block, and the Send Message block.

the motor block

Similar to the Move block, the *Motor block* controls the servo motors; but unlike the Move block, it specializes in precisely controlling one motor (Figure 8-12). In essence, the Motor block is more of a general-purpose block for controlling motors. Table 8-1 describes the symbols on the Motor block that show how it's configured.

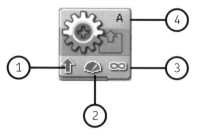

Figure 8-12: The Motor block

table 8-1: the motor block's symbols

symbol	purpose	default
1	Shows direction of motor rotation	Forward
2	Shows power level	75 percent
3	Shows unit of measure for duration	Unlimited
4	Shows selected output port	Output port A

In the Motor block's configuration panel, which is shown in Figure 8-13, there are eight parameters to configure, as well as a feedback box on the far left. This feedback box functions like the one on the Move block: It actively measures in degrees the movements of the selected motor when the NXT is connected to the computer. In fact, most of the Motor block's parameters function just as those for the Move block do. Here are the eight parameters:

Port Specify the output port (motor) to control: A, B, or C. You can select only one.

Direction Specify the direction of rotation for the motor: forward, reverse, or stop.

Figure 8-13: The Motor block's configuration panel

Action Specify how the motor accelerates by choosing Constant, Ramp Up, or Ramp Down from the drop-down menu. The *Constant* option immediately accelerates the motor to the power level specified in the Power parameter (the most common method). The *Ramp Up* option causes the motor to slowly accelerate to the specified power level. The *Ramp Down* option causes the motor to slowly decelerate to the specified power level.

NOTE If you choose Ramp Down in the Action parameter, the Motor block must be preceded by at least one other Motor block that sets the motor's shaft in motion and has its Next Action parameter set to Coast.

Power Specify the power level for the specified motor.

Control Specify whether to turn on *power control*. When the motor's shaft slows down due to slippage or resistance, this option increases the motor's power level up to 100 percent, attempting to keep the number of rotations per second consistent.

Duration Specify the duration for which the motor should run in seconds, degrees, or rotations; you can also choose unlimited.

Wait Specify whether the block should wait until it has completed before allowing the next block on the sequence beam to execute.

Next Action Specify whether the motors should brake or coast when finished.

NOTE When working with programs that control motors, it's good programming practice to use either the Move block or the Motor block but not both. Using only one of these two types of blocks decreases the size of the program file because all of the blocks of that type can share code.

the sound block

The Sound block in the Action blocks category is the same one in the Common blocks category. For more information about this block, consult "The Sound Block" on page 82.

the display block

The Display block in the Action blocks category is the same one in the Common blocks category. For more information about this block, consult "The Display Block" on page 84.

the send message block

The *Send Message block* sends a wireless message formatted as text, number, or logic data to another NXT via Bluetooth (Figure 8-14). This programming block is the key of NXT-to-NXT communication. Table 8-2 describes the symbols on the Send Message block that show how it's configured.

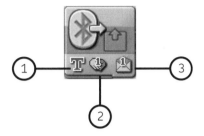

Figure 8-14: The Send Message block

table 8-2: the send message block's symbols

symbol	purpose	default
1	Shows message data type	Text
2	Shows connection number	1
3	Shows mailbox number	1

Before two NXTs can send or receive messages via Bluetooth, you must establish a connection between them, making one the master NXT and the other a slave NXT. The master NXT can, in fact, communicate with up to three slave NXTs, and each slave NXT communicates with the master NXT. Consult page 36 of your LEGO MINDSTORMS user guide or the Send Message block's documentation on the software for instructions on how to connect NXTs. Basically, the process consists of using the Bluetooth submenu on an NXT to initiate a search for other Bluetooth devices, selecting an NXT from the menu after the search, and entering passkeys to confirm the connection. The NXT you use to initiate the search will be the master NXT.

NOTE Remember that the Visible option on an NXT must be selected in order for other Bluetooth devices to detect the NXT. To access this option, select the Bluetooth submenu and then the Visibility subfolder.

Figure 8-15 shows the default configuration panel for the Send Message block. There are three parameters to configure:

Connection Specify the connection to which the block should send the message: 0, 1, 2, or 3. When you connect NXTs, each one has its own connection number, and the master NXT will always have connection 0. Sending a message to connection 1 (the default) will send a message to the first slave NXT.

Message Specify the format of the message—text, number, or logic data—and the message content.

Mailbox Specify the mailbox to which the block should send the message. Each NXT has ten mailboxes, and each mailbox can hold up to five messages. On the receiving end, the NXT can check a particular mailbox for a message.

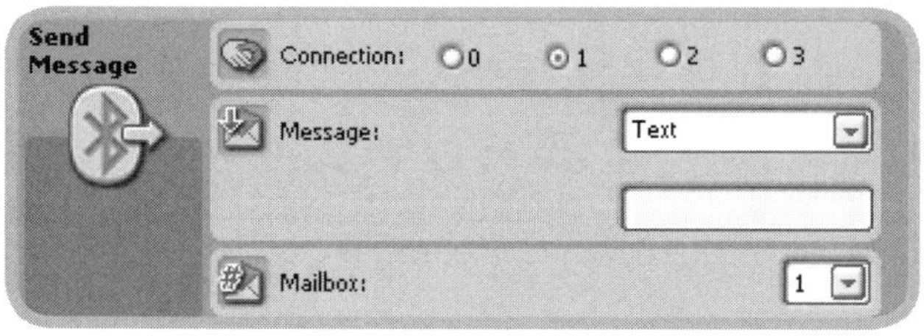

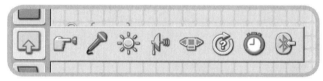

Figure 8-15: The Send Message block's configuration panel

Figure 8-16 shows an example program for a master NXT, and Figure 8-17 shows an example program for a slave NXT. In Figure 8-16, pressing the orange Enter button on the NXT sends a Bluetooth message to the slave NXT (mailbox 1) formatted as logic data with a value of *true*. After waiting a second, the master NXT then says, "Goodbye." In Figure 8-17, the program waits to receive the logic value *true* in its first mailbox. When it receives the correct message, the slave NXT says, "Hello." Since the master NXT says "Goodbye" just after the slave NXT says "Hello," the NXTs appear to be talking to each other!

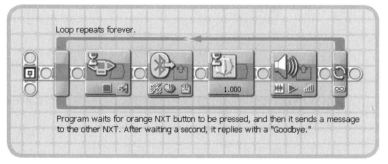

Figure 8-16: A master NXT program

Figure 8-17: A slave NXT program

the sensor blocks

The *Sensor blocks* read a sensor and send either the reading or other related data through a data wire to another block. For this reason, all Sensor blocks must use a data wire; however, clicking their data hub tabs reveals a number of optional data plugs that you can use to dynamically configure them. Positioning your cursor over the Sensor blocks category on the Complete palette displays a sub-palette with eight Sensor blocks: the Touch Sensor block, Sound Sensor block, Light Sensor block, Ultrasonic Sensor block, NXT Buttons Sensor block, Rotation Sensor block, Timer block, and Receive Message block (Figure 8-18).

Figure 8-18: The Sensor block's sub-palette includes eight sensor blocks.

Although the Wait, Loop, and Switch blocks discussed in the previous chapter can read all the same sensors, the Sensor blocks give us direct access to the sensor data. We commonly process sensor data with Data blocks, which we'll see examples of later in this chapter. All of the Sensor blocks behave similarly and are individually described in the software's NXT-G documentation, so we'll look at an example of using just one Sensor block: the Sound Sensor block.

The *Sound Sensor block* reads a sound sensor and can also test the value to see if it's greater than or less than a specified trigger point (Figure 8-19); Table 8-3 describes the symbols on the Sound Sensor block that show how it's configured. By default, the data hub only shows the *Sound Level plug*, which sends number data ranging from 0 to 100 (the possible sensor readings in percent). Clicking the data hub tab will display the rest of the available data plugs.

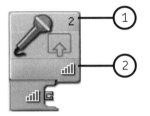

Figure 8-19: The Sound Sensor block

table 8-3: the sound sensor block's symbols

symbol	purpose	default
1	Shows selected input port	2
2	Shows the sound level at which the trigger point is set	More than 50

The configuration panel for the Sound Sensor block is shown in Figure 8-20. There are two parameters to configure, as well as a feedback box:

Port   Specify the input port on the NXT used by the sound sensor. You can select from ports 1, 2, 3, and 4.

Compare   Specify the trigger point by using the slider or typing a value in the text box. In addition, you can use the small drop-down menu or the radio buttons to specify whether the volume should be less than or greater than the trigger point. When this condition has been met, the Yes/No data plug (not shown) sends the *true* value; otherwise, it sends the *false* value. If you're not going to use this data plug, you can simply ignore this parameter.

Feedback Box   The feedback box on the left displays the current reading of the sound sensor (0–100) when the NXT is connected to the computer *and* the sound sensor is connected to the specified input port.

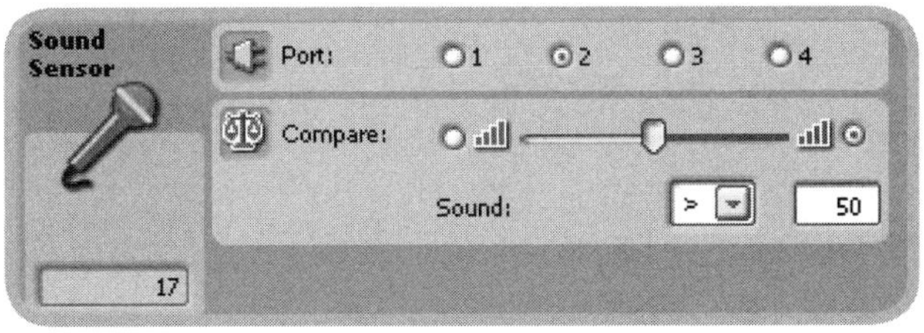

Figure 8-20: The Sound Sensor block's configuration panel

Figure 8-21 shows a program that works well with a robot like TriBot. In a Loop block set to repeat forever, a Sound Sensor block continuously sends its reading (0–100) to a Move block. The data wire connects to the Power plug on the Move block, causing the motors' power to change based on the sound sensor's reading. This means that noise, like a verbal command, causes the robot to move forward!

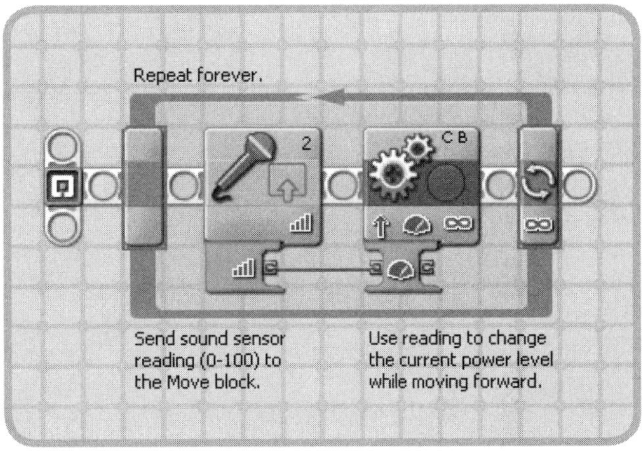

Figure 8-21: This program uses a Sound Sensor block to control the power level of a Move block.

the flow blocks

The Flow blocks control the flow of a program. Positioning your cursor over the Flow block category on the Complete palette displays a sub-palette with four Flow blocks: the Wait block, Loop block, Switch block, and Stop block (Figure 8-22). Since we already looked at the first three Flow blocks (see "The Common Palette" on page 79), we'll examine only the Stop block in this chapter.

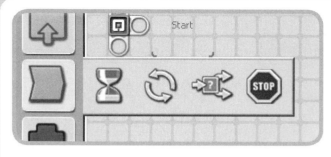

Figure 8-22: The Flow block's sub-palette includes the Wait block, Loop block, Switch block, and Stop block.

The *Stop block* is one of the simplest programming blocks of all: When executed, it stops the program (Figure 8-23). Any running motors will coast to a stop, and any sounds will be stopped.* The Stop block has no configuration panel, but it does have a data hub with an optional logic plug, the *Stop plug*. If the input value is *true*, the block stops the program; if the input value is *false*, the block doesn't stop the program.

Figure 8-23:
The Stop block

the data blocks

The *Data blocks* process data in a variety of ways and, like the Sensor blocks, send their output through a data wire to another block. Positioning your cursor over the Data block category on the Complete palette displays a sub-palette with six Data blocks: the Logic block, Math block, Compare block, Range block, Random block, and Variable block (Figure 8-24).

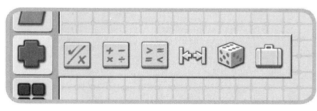

Figure 8-24: The Data block's sub-palette includes six data blocks.

---

* The Stop block also stops LEGO lamps, which aren't included in the NXT set. For more information about the LEGO lamp, consult the "Lamp* Block" topic on the software's documentation.

the logic block

The *Logic block* processes logic data (the values *true* and *false*) by performing a logical operation on two input values either set in the configuration panel or supplied by data wires (Figure 8-25). The output (also the value *true* or *false*), is sent through the bottom output plug. The symbol on the block uses a Venn diagram to illustrate which logical operation the block will perform.

Figure 8-25:
The Logic block

Figure 8-26 shows the default configuration panel for the Logic block, which has only one parameter: Operation. In the bottom portion of the panel, you can specify input values A and B if you're not going to do so with data wires. In the top portion of the panel, you can use the drop-down menu to select from four logical operations: And, Or, Xor, and Not.

And If both input values are true, the output is *true*. In all other cases, the output is *false*.

Or If one or both input values are true, the output is *true*. If both input values are false, the output is *false*.

Xor If one input value is true and the other is false, the output is *true*. If both values are true or both are false, the output is *false*.

Not Accepts only one input value and inverts it. For example, if the input value is false, the output is *true*.

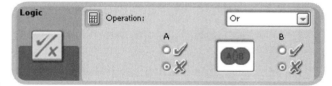

Figure 8-26: The Logic block's configuration panel

Figure 8-27 shows a practical application of the Logic block in a program. First, a Light Sensor block sends logic data to the Logic block with the value being *true* if it reads more than 75 percent and *false* if it doesn't. Then a Touch Sensor block sends logic data to a Logic block with the value being *true* if the touch sensor is pressed and *false* if it's not. The Logic block then performs an And operation on the data, which asks if both pieces of logic data are true.

Finally, the Move block receives the output data from the Logic block on its Steering plug and essentially says, *If this input value is true, output A goes in the forward direction;*

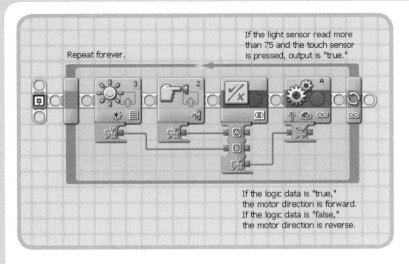

Repeat forever.

If the light sensor read more than 75 and the touch sensor is pressed, output is "true."

If the logic data is "true," the motor direction is forward. If the logic data is "false," the motor direction is reverse.

Figure 8-27: This program uses a Logic block to determine if a light sensor reads more than 75 percent at the same time that a touch sensor is pressed.

if this input value is false, output A goes in the reverse direction. All of the blocks are in a Loop block set to repeat forever so that the test will continuously be performed. Therefore, holding the light sensor up to a bright enough light source and pressing the touch sensor at the same time causes the motor on port A to rotate in the forward direction. Otherwise, it'll rotate in the reverse direction.

the math block

The *Math block* performs simple calculations with number data using addition, subtraction, multiplication, or division (Figure 8-28). The symbol on the front of the block shows which operation the block will perform (addition by default). The calculation involves two input values (A and B) that you either type in the block's configuration panel or supply through data wires.

Figure 8-28:
The Math block

The Math block's configuration panel has only the Operation

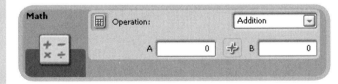

Figure 8-29: The Math block's configuration panel

parameter (Figure 8-29). The drop-down menu enables you to select Addition, Subtraction, Multiplication, or Division. If you're not going to supply the input values A or B through data wires, you can type values in the A and B text boxes.

In Figure 8-30 a program uses an Ultrasonic Sensor block to send the readings of an ultrasonic sensor to a Math block, which multiplies the number data by two (specified in its configuration panel). The Math block's output then goes to the Power plug on a Motor block. Therefore, the motor's power will always be double what the ultrasonic sensor reads. For instance, if the ultrasonic sensor reads 40 inches, then the Motor block's Power parameter would be 80 percent.

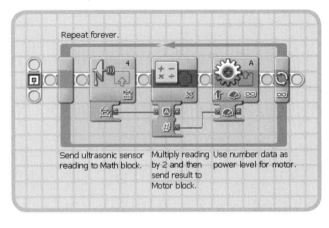

Repeat forever.

Send ultrasonic sensor reading to Math block.

Multiply reading by 2 and then send result to Motor block.

Use number data as power level for motor.

Figure 8-30: This program uses a Math block to multiply an ultrasonic sensor's reading by two, and then it sends the output to a Motor block.

the compare block

The *Compare block* performs comparisons on number data (Figure 8-31). Using input values either set in the configuration panel or supplied by data wires, it compares two values and sends the output (either *true* or *false*) through a data wire. The symbol on the front of the block shows the type of comparison the block will perform (*Less than* by default).

Figure 8-31:
The Compare block

The configuration panel has one parameter (Operation) with three options (Figure 8-32). If you're not going to supply the A and B input values through data wires, you can type them in the A and B text boxes. The drop-down menu enables you to select Less Than, Greater Than, or Equals. If you select Less Than, the block determines whether input value A is less than input value B. If you select Greater Than, the block determines whether input value A is greater than input value B. If you select Equals, the block determines if input value A is equal to input value B. In each case, if the comparison is true, the block's output is *true*; if the comparison isn't true, the block's output is *false*.

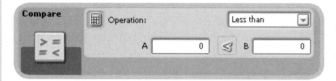

Figure 8-32: The Compare block's configuration panel

In Figure 8-33, a program uses a Sound Sensor block to send the reading of a sound sensor to a Compare block for an A input value. The B input value on the Compare block is set to 50 in its configuration panel. The Compare block asks, *Is this input value (the sensor reading) greater than 50?* If it is, the Compare block sends a value of *true* to the Motor block; if it isn't, the Compare block sends a value of *false* to the Motor block. Therefore, if a sound sensor on input port 2 reads a sound level greater than 50 percent, a motor on output port A will rotate in the forward direction. Otherwise, the motor rotates in the reverse direction.

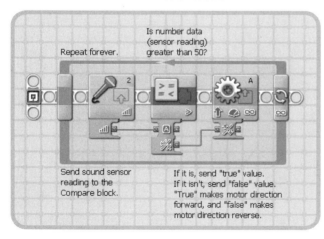

Figure 8-33: This program uses a Compare block to determine whether a sound sensor reads a value greater than 50, and then it sends the output to the Direction plug on the Motor block.

the range block

The *Range block* is basically a Compare block that specializes in testing a value to see if it's inside or outside a range of specified numbers (Figure 8-34). If the test value meets the criterion, the output value is *true*. If the test value doesn't meet the criterion, the output value is *false*. The symbol on the front of the Range block shows whether the block is testing to see if a value is inside or outside of a range (inside range by default).

Figure 8-34:
The Range block

The Range block's configuration panel (Figure 8-35) has two parameters: Operation and Test Value. In the Operation parameter, a drop-down menu gives you two options: Inside Range and Outside Range. These specify whether you're testing for a value inside a range of numbers or outside a range of numbers. In either case, you can set the upper and lower boundary numbers by moving the sliders, typing values in the A and B text boxes, or using data wires on the A and B data plugs. Similarly, you can type the test value in the text box in the Test Value parameter or by using a data wire on the Test Value plug. With the panel configured as shown, the Range block would ask, *Is 50 between 25 and 75, or equal to either boundary number?*

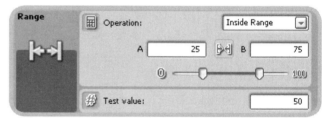

Figure 8-35: The Range block's configuration panel

the random block

The *Random block* (Figure 8-36) is the simplest Data block, but it's a lot of fun to use! You can set the minimum and maximum limits for the random number through the block's input plugs (not shown) or in its configuration panel. With the panel configured as shown in

Figure 8-36:
The Random block

Figure 8-37, the Random block would produce a value somewhere between 0 and 100, inclusive. The output (random number) is sent through a data wire from the block's output plug.

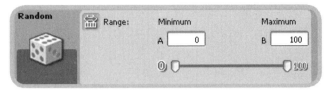

Figure 8-37: The Random block's configuration panel

Figure 8-38 shows a program that uses three Random blocks to dynamically control a Sound block. Each Random block sends number data to a different plug on the Sound block to control the duration, volume, and frequency of the tone that the Sound block plays. All the blocks are in a Loop block set to repeat forever, so the robot will keep making music or, more likely, just noise!

the variable block

The *Variable block* writes number, text, or logic data to *variables*—locations in the NXT's memory—and reads data from variables (Figure 8-39). Table 8-4 describes the symbols on the Variable block and its data hub that show how the block is configured. This block is useful for storing and retrieving data that a robot may gather, produce, or receive.

Before a Variable block can use a variable, you must first define it. Each variable has its own name, a data type,

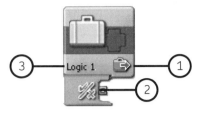

Figure 8-39: The Variable block

table 8-4: the variable block's symbols

symbol	purpose	default
1	Shows action	Read
2	Shows data type	Logic
3	Shows variable name	Logic 1

and a value. Selecting Edit ▸ Define Variables from the menu bar displays the Edit Variables dialog (Figure 8-40). There are always three default variables that you can use: *Logic 1* (for logic data), *Number 1* (for number data), and *Text 1* (for text data). If you want to create your own variable, click the Create button to add another variable to the list. You can then give the variable a name and specify its data type.

NOTE When you define a variable, it's available only for the active program. You must define each program's own variables unless you're using the default variables.

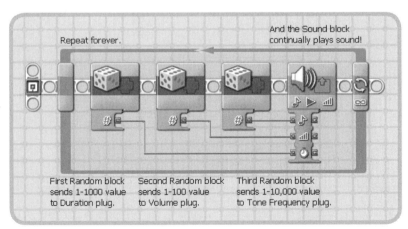

First Random block sends 1-1000 value to Duration plug.

Second Random block sends 1-100 value to Volume plug.

Third Random block sends 1-10,000 value to Tone Frequency plug.

Repeat forever.

And the Sound block continually plays sound!

Figure 8-38: This program uses three Random blocks to control the duration, volume, and frequency of a tone played by a Sound block.

Figure 8-40: Use the Edit Variables dialog to define new variables for your program.

After defining one or more variables, you can select them from the configuration panel of any Variable block within the program (Figure 8-41). There are three parameters to configure:

List Select the variable you want to use. Besides the three default variables, any variables you've created for your program will also appear here.

Action Specify whether to read data (retrieve existing data from a variable) or write data (add new data to a variable). Selecting Read causes the data hub to show a single output plug to which you can attach a data wire and from which you can transmit the data to another block. Selecting Write causes the data hub to show an input plug with a corresponding output plug. Writing data allows you to either supply the input values through the data hub or specify the content through the Value parameter.

NOTE Be careful! Writing data overwrites any existing data in the variable.

Value When the Action parameter is set to Write, you can specify the variable's contents in this parameter.

Figure 8-41: The Variable block's configuration panel

In Figure 8-42, a program uses a Variable block and the default Number 1 variable to control the power level of a Motor block. Every time you "bump" (press and release) a touch sensor, the blocks on the lower sequence beam increment the Number 1 variable by 10. To accomplish this, a Math block reads the Number 1 variable, adds 10 to the value, and then writes the output to the Number 1 variable. On the upper sequence beam, a Motor block continuously reads the Number 1 variable for its Power plug. Therefore, each time you bump the touch sensor, the motor's power increases by 10 percent.

Continuously use the value of variable "Number 1" for Power level.

When the touch sensor is bumped... ...a Math block reads the Number 1 variable and adds 10 to it.

The Number 1 variable then writes the output of the Math block to its value.

Repeat forever.

Figure 8-42: This program uses the Variable block and the Number 1 variable to control the power level of a Motor block.

the advanced blocks

The *Advanced blocks* perform miscellaneous advanced functions. Positioning your cursor over the Advanced block category on the Complete palette displays a sub-palette with six Advanced blocks: the Text block, Number to Text block, Keep Alive block, File Access block, Calibration block, and Reset Motor block (Figure 8-43). As we individually examine these blocks, remember that you can configure most of their options not only in their configuration panels, but also with data wires.

Figure 8-43: The Advanced block's sub-palette includes six advanced blocks.

the text block

The *Text block* combines up to three separate pieces of text data into one piece of text (Figure 8-44). Using either the input plugs on the data hub or the block's configuration panel

(Figure 8-45), you can specify the text to add. You then use the last output plug to send the output to another block. Be careful to include spaces in the individual pieces of text data where appropriate. For example, using the Text block to add "The", "Text", and "Block" would result in *TheTextBlock*. But adding "The ", "Text ", and "Block" would result in *The Text Block*.

Figure 8-44:
The Text block

Figure 8-45: The Text block's configuration panel

the number to text block

The *Number to Text block* converts number data into text data so that you can display it on the LCD (Figure 8-46). You can specify the number to convert by either supplying number data with a data wire or typing a number in the block's configuration panel (Figure 8-47).

Figure 8-46:
The Number to Text
block

Figure 8-47: The Number to Text block's configuration panel

A practical application of the Number to Text block involves converting a sensor reading into text data and then displaying the text data on the LCD. Figure 8-48 shows a

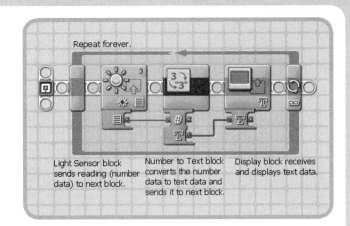
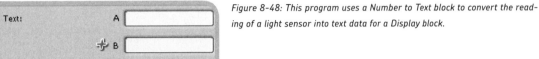

Repeat forever.

Light Sensor block sends reading (number data) to next block.

Number to Text block converts the number data to text data and sends it to next block.

Display block receives and displays text data.

Figure 8-48: This program uses a Number to Text block to convert the reading of a light sensor into text data for a Display block.

program that continuously converts a light sensor's reading (number data) into text data with a Number to Text block and then transmits the data to the Text plug on a Display block. As the light sensor "sees" areas of varying lightness, the LCD would update the reading.

the keep alive block

The *Keep Alive block* prevents the NXT from "falling asleep" by overriding the Sleep mode settings configured on the Settings sub-menu of the NXT (Figure 8-49). There is no configuration panel for this block and only one optional data plug (see the block's data hub chart).

Figure 8-49:
The Keep Alive block

the file access block

The *File Access block* stores and reads text or number data files on the NXT (Figure 8-50). Table 8-5 describes the symbols on the File Access block that show how it's configured. While very similar to the Variable block, the File Access block has the advantage of retaining data stored on the NXT even when you exit a program or turn off the NXT. The disadvantage to using the File Access block, however, is that it's more difficult to use and takes up more of the NXT's memory.

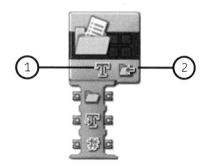

Figure 8-50: The File Access block

table 8-5: the file access block's symbols

symbol	purpose	default
1	Shows data type	Text
2	Shows action	Write

Figure 8-51 shows the File Access block's configuration panel, which has options similar to ones on the Variable

block and even the Record/Play block. Here are the five parameters:

Action   Specify whether the block writes data to a file, reads a data file, deletes a data file, or closes a data file. The writing and reading actions for this block are essentially the same as the writing and reading actions for the Variable block. Since writing to a data file *doesn't* overwrite any existing data, however, you use the Delete option to clear a file's contents. Importantly, you must use the Close option to close a data file whenever you perform a different action on it. For example, between a File Access block that reads a data file and a File Access block that writes to the same data file, you would need to include a File Access block that closes the data file.

Name   Specify a name for the data file that the Action parameter will create or use.

File   If the NXT is connected to your computer, you can use this parameter to select from existing files on the NXT.

Type   If reading or writing, specify the type of data involved: number or text.

Text/Number   If reading or writing, type the data for the data file (unless you're supplying it with data wires).

Figure 8-51: The File Access block's configuration panel

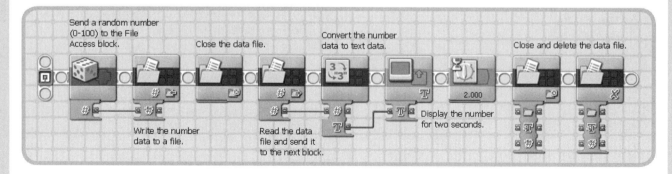

Figure 8-52: This program uses the File Access block to store and read data from a Random block.

Figure 8-52 shows a program that stores a random number with a File Access block and then displays that number on the LCD. The program begins with a Random block that sends a number to a File Access block that writes that number to a data file. Next, another File Access block closes the data file. Another File Access block reads the data and then sends the data to a Number to Text block that converts it to number data. A Display block displays the number for two seconds with a Wait block, and then another File Access block closes the data file. Finally, a File Access block deletes the file. Notice that I had to close the data file each time I performed a different action on it.

NOTE The program in Figure 8-52 isn't an example of the most efficient way to display a random number; it's merely an example of how to use the File Access block.

the calibration block

The *Calibration block* calibrates a light sensor or sound sensor by setting the minimum (0 percent) or maximum (100 percent) value that the sensor can read (Figure 8-53).

Table 8-6 describes the symbols on the Calibration block that show how it's configured.

Figure 8-54 shows the configuration panel for the Calibration block. There are four parameters to configure:

Port   Specify which port the sensor is using on the NXT.

Sensor   Specify whether to calibrate a light or sound sensor.

Action   Specify whether you should calibrate a sensor or delete a sensor (i.e., return a sensor to the default settings).

Value   Specify whether the calibration should be for the maximum or the minimum value.

You must use one Calibration block for calibrating the maximum value and one for calibrating the minimum value, but you don't necessarily need to calibrate both. You might want to keep the default settings for one of the values. In addition, when the Calibration block within a program executes, make sure that you have the sound or light sensor ready to read what will be its maximum or minimum value.

Figure 8-53: The Calibration block

table 8-6: the calibration block's symbols

symbol	purpose	default
1	Shows type of sensor being calibrated	Light sensor
2	Shows whether minimum or maximum value is being set	Minimum

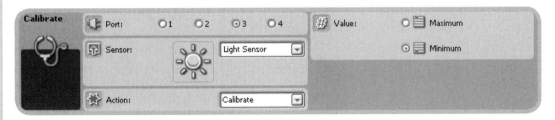

Figure 8-54: The Calibration block's configuration panel

NOTE If you don't necessarily need to calibrate a light and/or sound sensor within a program using a Calibration block, you can select Tools ▸ Calibrate Sensors from the menu bar to calibrate the sensors with the Calibrate Sensor dialog (make sure your NXT is connected first!). For more information on this feature, see the "Calibrate Sensors" topic on the software's documentation.

the reset motor block

The *Reset Motor block* resets an automatic error correction mechanism in the servo motors (Figure 8-55). The letters on the Reset Motor block show which output ports will be reset, and the configuration panel (Figure 8-56) has only the Port parameter that lets you select which output ports to reset.

Figure 8-55:

The Reset Motor block

Figure 8-56: The Reset Motor block's configuration panel

Why would you want to use the Reset Motor block? First, you must understand that when you set the Next Action parameter for Move or Motor blocks to Coast, the actual number of rotations will not *exactly* match the amount set in the Duration parameter. The automatic error correction mechanism normally corrects these errors, but you can use the Reset Motor block in between Move or Motor blocks to prevent it from doing so. You'll rarely (if ever) use this block, but you may find it helpful in certain circumstances (see the Reset Motor block's documentation on the software for detailed scenarios).

the custom palette

Finally, the only programming blocks not included in the Complete palette are those in the *Custom palette*, which contains *Custom blocks*. The Custom palette breaks the Custom blocks down into two subcategories: My Blocks and Web Blocks (Figure 8-57). Positioning your cursor over the My Blocks icon or the Web Blocks icon reveals a sub-palette with any blocks belonging to that subcategory. There are initially no programming blocks in the Custom palette, however. You must create Custom blocks!

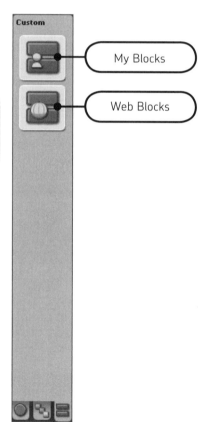

Figure 8-57: The Custom palette includes two sub-categories of blocks: My Blocks and Web Blocks.

the my blocks

The basis of My Blocks is the standard programming blocks. In other words, a *My Block* represents one or more standard blocks and executes those blocks when it's executed. We typically use My Blocks to group together a series of blocks configured for a specific action (e.g., grabbing a ball). Therefore, every time we want to instruct a robot in a program to perform that action, we can use a single My Block. Not only do My Blocks save us from having to insert many additional blocks into a program, but if we want to change an action, we can directly modify a My Block and any occurrences of that block in the program are automatically updated.

Creating a My Block is a simple process. Go to the work area and add a few blocks, as shown in Figure 8-58, and configure them any way you want. We'll use these blocks to make a simple My Block.

Figure 8-58: We'll use these four blocks as the basis of a My Block.

Next, select all the blocks and then choose **Edit ▸ Make a New My Block** from the menu bar or click the Create My Block icon ▤ on the toolbar. The My Block Builder dialog will appear, in which you can create the My Block (Figure 8-59). Type **Test Block** in the text box in the upper-right corner of the dialog to set that as the block's name. Optionally, you can type a block description beneath it. Then click the **Next** button in the lower-left corner of the dialog. In the second step, you can create a customized icon for your My Block by dragging icons into the white box near the top of the screen (Figure 8-60). When you're finished, click the **Finish** button.

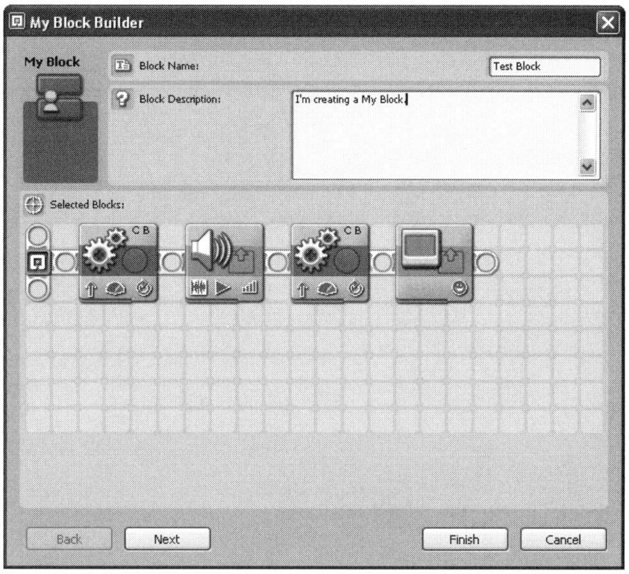

Figure 8-59: Step one in the My Block Builder dialog

Figure 8-60: Step two in the My Block Builder dialog

The work area should now look like Figure 8-61 with a single My Block (the icon on the block will depend on what you chose). To view or modify the underlying blocks, you can double-click the My Block. If you position your cursor over the My Block subcategory on the Custom palette, it should display a sub-palette that contains your new My Block. When you want to delete My Blocks that you've made, select **Edit ▶ Manage Custom Palette** from the menu bar. With a My Block selected in the work area, you can also select Edit ▶ Edit Selected My Block to modify a My Block's contents and Edit ▶ My Block Icon to change its icon.

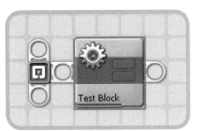

Figure 8-61: The finished My Block—Test Block

NOTE You can create My Blocks with functional data hubs by using blocks on only one end of a data wire. For example, if you have two blocks connected by a data wire, selecting *one* of them and making a My Block from it would result in a My Block with the appropriate type of data plug. Experimenting with this aspect of My Blocks will reveal the various types of data hubs you can achieve.

the web blocks

A *Web Block* is simply a My Block that someone else has created and that you've presumably downloaded from the Internet. To add these blocks to the Web Block sub-palette, select **Edit ▶ Manage Custom Palette** on the menu bar and navigate to the Web Blocks folder. (The NXT-G interface must be active to access this menu item.) Place any My Blocks you've downloaded into this folder, and they'll be added to the Web Blocks subcategory on the Custom palette.

NOTE A Web Blocks folder doesn't appear on your computer until you've first positioned your cursor over the Web Blocks subcategory on the Custom palette.

BROKEN MY BLOCKS

If a program uses a My Block that isn't in the Custom palette or some other expected location on your computer, the block becomes *broken* (Figure 8-62). Programs with broken My Blocks cannot be downloaded to an NXT, just as programs with broken data wires cannot be downloaded. The main cause of broken My Blocks is deleting a My Block from the Custom palette while an existing program still uses that block. Another cause is downloading an NXT-G program from the Internet that uses My Blocks but failing to download the My Blocks that go along with it. When you find a broken My Block in a program, you can delete it and then create a new My Block or download the necessary My Block from the Internet.

Figure 8-62:
A broken My Block

expanding NXT-G with the dynamic block update

The official MINDSTORMS website has a page devoted to software updates for the NXT set at http://mindstorms .lego.com/support/updates. Shortly after the release of the NXT set, the *Dynamic Block Update* was added to this page. Installing this update enables you to import new blocks to the Complete palette. Currently, the page offers two additional libraries of blocks that you can import: the Mini Block Library and the Legacy Block Library.

The *Mini Block Library* is a collection of blocks—the Mini Move, Mini Motor, Mini Display, and Mini Sound blocks—that are identical or very similar in functionality to the standard Move, Motor, Display, and Sound blocks but take up less memory on the NXT. You'll find the Mini Block Library useful if program size is important to you or if you're having difficulty fitting your programs onto your NXT. However, if you're using NXT-G 1.1, you do not need to download the Mini Block Library; NXT-G 1.1 offers the most efficient blocks available.

The *Legacy Block Library* is a collection of blocks—the Lamp block, Motor block, Touch Sensor block, Light Sensor block, Rotation Sensor block, and Temperature Sensor block—that work with the old MINDSTORMS electronic elements. You must use a special converter cable to connect these electronic pieces to the NXT; see the note in "The Input (Sensor) Ports" on page 22 for more information about this cable.

Finally, LabVIEW users can employ the LabVIEW Toolkit for LEGO MINDSTORMS NXT (http://www.ni.com/academic/ mindstorms) to develop new types of NXT-G blocks, and you can similarly use the Dynamic Block Update to import these blocks to the Complete palette. The popular blog at http://nxtasy.org has collected a number of these user-made blocks into a repository, which is available at http://nxtasy .org/category/nxt-repository/programming.

conclusion

Exploring advanced NXT-G concepts and programming blocks opens up new possibilities for our NXT creations. In this chapter, we first observed how to use data wires to transmit data between programming blocks in a variety of ways. After that our focus was the Complete palette, which contains all of the standard programming blocks. Next, we observed the Custom palette and how to make our own blocks. Finally, we saw how to use the Dynamic Block Update to import new blocks, such as the Mini Block Library. In the next chapter, we'll conclude Part III with a brief look at several unofficial text-based programming languages for the NXT.

unofficial programming languages for the NXT

While NXT-G provides a colorful and user-friendly approach to programming NXT robots, some people prefer programming with a text-based language, which involves writing lines of code rather than dragging and dropping code blocks. There are a variety of unofficial text-based languages for the NXT freely available on the Internet. Generally speaking, these programming languages offer greater control but can be more difficult to learn.

I'll introduce four important text-based programming languages for the NXT in this chapter—NBC, NXC, leJOS NXJ, and RobotC—and then I'll briefly discuss a powerful (and free!) program called BricxCC with which you can easily create and manage text-based programs. The goal of this chapter is not to teach you how to program in these unofficial languages but to familiarize you with text-based NXT programming in general. Tutorials, programmer guides, and sample programs are available on the languages' websites, and they can help you learn how to do the actual programming.

NOTE Appendix C lists all the websites mentioned throughout this chapter, as well as many additional websites that offer programming resources for the NXT.

NeXT byte codes

John Hansen developed the first text-based programming language for the NXT: *NeXT Byte Codes (NBC)*. Based on assembly language, NBC code is relatively simple and consists mainly of abbreviated and whole words. NBC uses the standard firmware on the NXT, so you don't have to change anything on your NXT to use NBC programs. The NBC software is available for free on the NBC website (http://bricxcc.sourceforge.net/nbc) and it runs on Windows, Mac OS X, and Linux machines.

The sample NBC code in Listing 9-1 works well with a simple mobile robot like TriBot: It drives the robot forward (using ports B and C) for five seconds with the motors synchronized, brakes the motors, drives the robot in reverse for five seconds with the motors synchronized, and finally stops. Comments explaining the code are preceded by // and continue to the end of the line.

After writing an NBC program, you use the NBC compiler to convert the code into byte codes for the NXT. (The NXT can interpret or "read" byte codes, but it can't read the actual NBC code.) After compiling the

```
// Include important constants and macros
#include "NXTDefs.h"

// Begin main thread
thread main
  OnFwdReg(OUT_BC, 75, OUT_REGMODE_SYNC)  // Power motors forward
  wait 5000                               // Wait five seconds
  Off(OUT_BC)                             // Brake motors
  OnRevReg(OUT_BC, 75, OUT_REGMODE_SYNC)  // Power motors in reverse
  wait 5000                               // Wait five seconds
  Off(OUT_BC)                             // Brake motors
endt   // End main thread
```

Listing 9-1: Some sample NBC code

program, you then download it to your NXT. Amazingly, NBC programs on the NXT can be up to ten times smaller and faster than NXT-G programs that perform the same tasks!

not eXactly C

Shortly after creating NBC, Hansen developed the high-level language *Not eXactly C (NXC)*. As its name suggests, NXC is based on the C programming language. Furthermore, NXC code is very similar to Not Quite C (NQC) code. Originally created by Dave Baum and then later maintained by Hansen, NQC is the extremely popular programming language for the MINDSTORMS RCX microcomputer. If you're familiar with NQC, making the transition to NXC will be easy.

Notably, you also use the NBC compiler for NXC programs. In fact, the NBC compiler converts NXC code to NBC code and *then* compiles it. Because of their close relation (i.e., NXC is built on top of the NBC compiler), NBC and NXC share the same website, which is available at http://bricxcc .sourceforge.net/nbc.

The sample NXC code in Listing 9-2 also works well with a simple mobile robot like TriBot and uses a touch sensor on input port 1 to control the driving motors on output ports B and C. In short, this code instructs the robot to drive forward at a power level of 75 percent until the touch sensor is pressed, drive in reverse until the touch sensor is released, and finally stop after waiting half a second. The code also synchronizes the motors to ensure that the robot drives in a straight line. Again, comments explaining the code are preceded by // and continue to the end of the line.

In summary, NBC and NXC are excellent languages for those who have no experience programming with text-based languages (or any programming experience, for that matter), but they still offer features to satisfy more advanced programmers. In addition, despite the fact that the sample code performs tasks that could easily be accomplished with NXT-G, NBC and NXC have many capabilities that NXT-G doesn't. Visiting the NBC/NXC website is the best place to begin learning how to use these languages.

leJOS NXJ

Given the popularity of the Java programming language, it's no surprise that it has been ported to MINDSTORMS microcomputers—at least in the form of leJOS. The first LEGO microcomputer to use leJOS was the RCX, but with the release of the NXT set, the leJOS team released a version for the NXT known as *leJOS NXJ*. You can download the latest version of leJOS NXJ for free from the leJOS website at http://lejos.sourceforge.net; the software runs on Windows, Linux, and, in the near future, Mac OS X machines.

NOTE The word *leJOS* means *the Java Operating System* and is a play on the word *LEGOs*. According to leJOS documentation, you pronounce *leJOS* like the Spanish word *lejos*, which means *far*.

```
// Include important constants and macros
#include "NXCDefs.h"

// Begin main task
task main()
{
  // Initialize touch sensor on input port 1
  SetSensorTouch(S1);

  // Power motors B and C forward while synchronized at power level 75
  OnFwdReg(OUT_BC, 75, OUT_REGMODE_SYNC);
  until(Sensor(S1) == 1);    // Do this until touch sensor is pressed

  // Power motors B and C in reverse while synchronized at power level 75
  OnRevReg(OUT_BC, 75, OUT_REGMODE_SYNC);
  until(Sensor(S1) == 0);   // Do this until touch sensor is released

  Wait(500);         // Wait for half a second
  Off(OUT_BC);       // Brake motors B and C
}
```

Listing 9-2: Some sample NXC code

Unlike NBC and NXC, the leJOS NXJ programming language requires you to use custom firmware on the NXT to execute the code. Instead of calling it *firmware*, however, we refer to it as a *Java Virtual Machine (JVM)*. Even though you must replace the standard firmware on your NXT with the JVM, you can always re-install the standard firmware later.

At the most basic level, leJOS NXJ truly is Java, so you'll need a working knowledge of Java to successfully use this language. In addition, setting up leJOS NXJ isn't as easy as setting up NBC or NXC. Once you've gotten past these initial steps, though, you'll discover the incredible power of leJOS NXJ and its many unique features. An excellent way to learn leJOS NXJ is with Brian Bagnall's *Maximum LEGO NXT: Building Robots with Java Brains* (Variant Press, 2007).

The code in Listing 9-3 is taken from the leJOS website. Most of the action happens in the **for** loop that repeats endlessly, displaying the reading of each motor's tachometer (built-in rotation sensor) and the raw value of a touch sensor on the NXT's LCD. It also causes the motor on port A to start and stop when you press the touch sensor.

```
import lejos.nxt.*;

public class TestMotor
{
  public static void main (String[] aArg)
  throws Exception
  {
    String m1 = "Motor A: ";
    String m2 = "Motor B: ";
    String m3 = "Motor C: ";
    String p = "Port S1: ";

    TouchSensor touch = new TouchSensor(Port.S1);

    for(;;) {
      LCD.clear();
      LCD.drawString(m1,0,1);
      LCD.drawInt(Motor.A.getTachoCount(),9,1);
      LCD.drawString(m2,0,2);
      LCD.drawInt(Motor.B.getTachoCount(),9,2);
      LCD.drawString(m3,0,3);
      LCD.drawInt(Motor.C.getTachoCount(),9,3);
      LCD.drawString(p,0,4);
      LCD.drawInt(Port.S1.readRawValue(),9,4);
      Motor.A.setSpeed(100);
      if (touch.isPressed()) Motor.A.forward();
      Thread.sleep(1000);
      if (touch.isPressed()) Motor.A.stop();
      LCD.refresh();
      Thread.sleep(1000);
    }
  }
}
```

Listing 9-3: Some sample leJOS NXJ code

RobotC

The *RobotC* programming language is a commercial product that targets the educational market and supports not only the NXT but also other microcomputers. Developed by Carnegie Mellon University's Robotics Academy, RobotC uses an almost full implementation of C, unlike NXC which is only *like* C. RobotC is also another firmware replacement language, so your NXT will need to use custom RobotC firmware to execute RobotC programs.

On the RobotC website, which is available at http://www.robotc.net, you can purchase one or more user licenses. To purchase a single user license and immediately download the software currently costs $30. You can also download RobotC and try it for 30 days before purchasing a user license. As shown in Figure 9-1, the central feature of the RobotC software is its powerful *integrated development environment (IDE)* with which you can write, compile, and download your RobotC programs. The software runs only on Windows XP, but reportedly will work with Mac platforms in the future.

The RobotC firmware has numerous enhancements over the standard firmware, providing faster-executing code, better memory management, expanded sound capabilities, and much more. If you need help with RobotC, you have access to a full-time support staff, extensive online help material, and many other resources. Overall, RobotC is suitable for both beginning and advanced programmers, and it works well not only for educators but also for MINDSTORMS hobbyists seeking a text-based language for the NXT or another microcomputer.

The code in Listing 9-4 is taken from a sample program (nxt_line_track.c) included with the RobotC software, and it instructs a simple mobile robot to follow a line in reverse. In a loop that repeats forever, the program continually reverses one motor and stops the other based on the reading of a light sensor. Once again, comments are preceded by // and continue to the end of the line. You should recognize other similarities to the NXC code shown earlier as well.

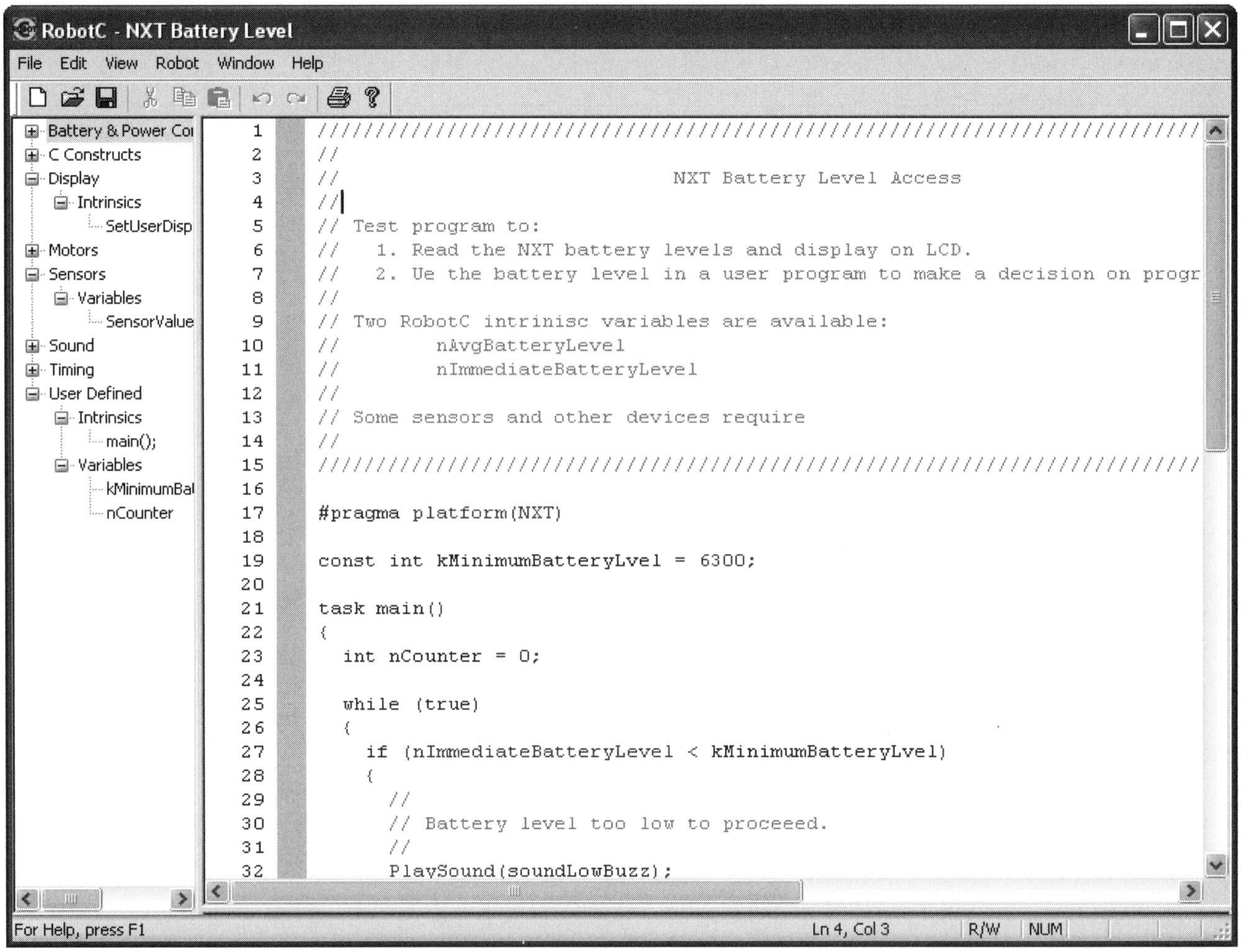

Figure 9-1: The RobotC IDE

```
const tSensors lightSensor = (tSensors) S1  // sensorLightActive
task main()
{
   while(true) // Infinite while loop declared with "true" condition
   {
      // If light sensor reads value less than 45, if code will be run
      if(SensorValue(lightSensor) < 45)
      {
       motor[motorA] = -75; // Motor A is reversed at a -75 power level
       motor[motorB] = 0;   // Motor B is stopped with a 0 power level
      }
      else
      {
       motor[motorA] = 0;  // Motor A is stopped with a 0 power level
       motor[motorB] = -75;  // Motor B is reversed at a -75 power level
      }
   }
}
```

Listing 9-4: Some sample RobotC code

bricx command center

If you're using a text-based programming language that doesn't have its own specific IDE, how do you write, compile, and download your programs to the NXT? One way is to write code in a simple text editor program and then compile it and download it to the NXT from a command prompt. But a much more efficient way is to use an IDE like *Brick Command Center (BricxCC)*. The main feature of BricxCC is a programmer's editor in which you write your programs.

Figure 9-2 shows BricxCC version 3.3.7.16 with an NXC file open in the programmer's editor. BricxCC is available for free at http://bricxcc.sourceforge.net, but it only supports Windows (most versions).

Because John Hansen maintains BricxCC, NBC, and NXC, BricxCC is especially suited to NBC/NXC programming. In fact, BricxCC now includes the latest versions of NBC and NXC. However, BricxCC supports a large variety of other languages, including Java (i.e., leJOS). Accessing the Preferences dialog in BricxCC enables you to configure how BricxCC handles these languages (e.g., how it compiles code for a given language), but you should also check the BricxCC website for instructions about properly setting up different languages.

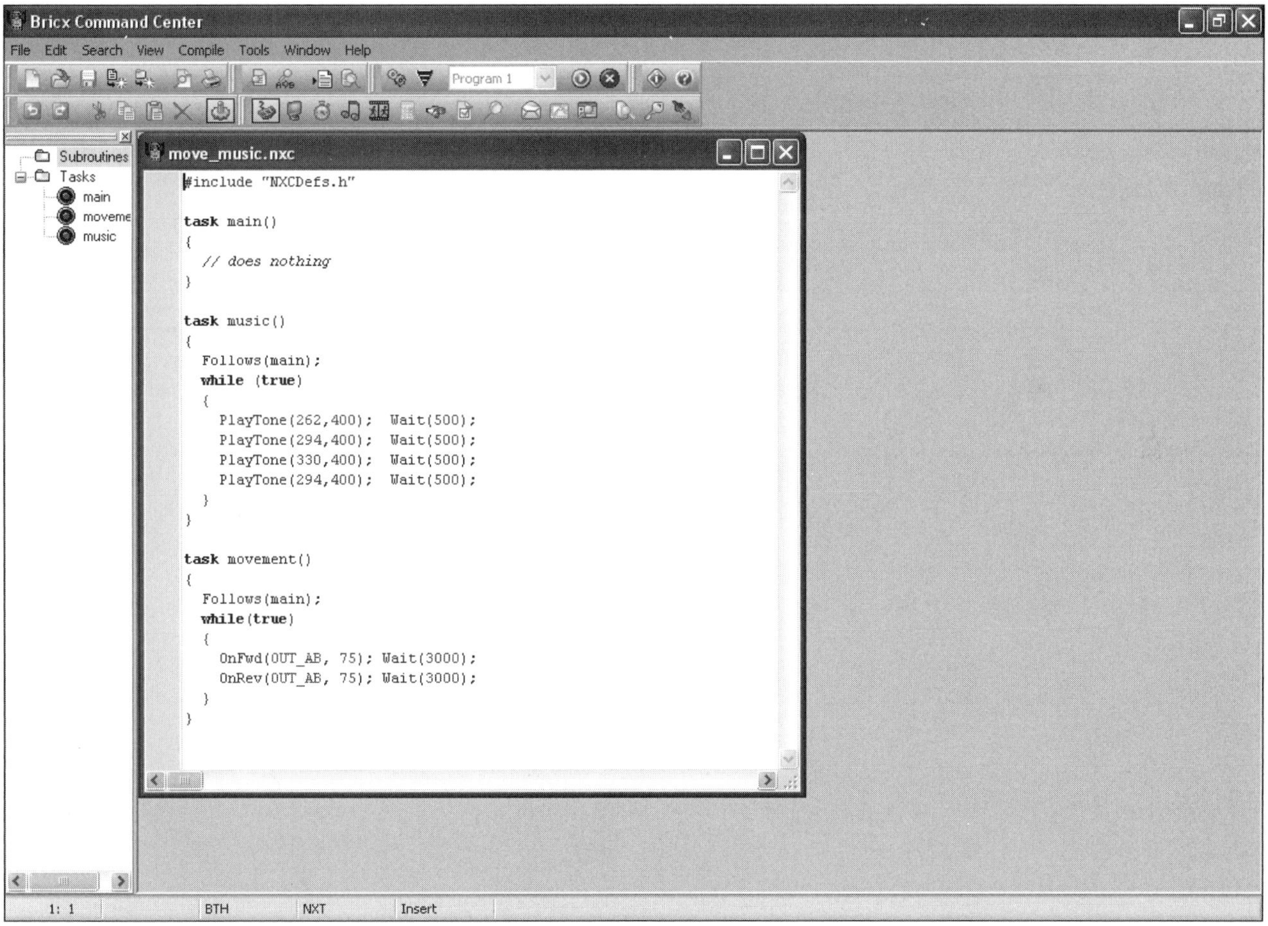

Figure 9-2: The BricxCC IDE

summarizing the programming languages

The four programming languages we've examined are different in a number of respects, and Table 9-1 summarizes their characteristics. You can use this data to determine which programming language is best for you based on factors such as the type of computer you're using and your level of programming experience.

conclusion

Programming for the NXT extends much further than the official NXT-G programming language. In this chapter, we observed four text-based programming languages that are completely unofficial but very powerful: NBC, NXC, leJOS NXJ, and RobotC. Each of these languages offers unique features and capabilities to assist you in creating programs that make your robots do exactly what you want them to do.

You've reached the end of Part III and should now have a solid understanding of NXT programming. In Part IV, you'll create complete NXT robots and use many of the building and programming techniques you've learned—and discover some new ones as well!

table 9-1: text-based programming languages for the NXT

language	free?	operating system(s)	firmware replacement?	difficulty	based on
NBC	Yes	Windows, Mac OS X, and Linux	No	Low	Assembly language
NXC	Yes	Windows, Mac OS X, and Linux	No	Low	C
leJOS NXJ	Yes	Windows and Linux (support for Mac OS X coming soon)	Yes	Moderate to high	Java
RobotC	No	Windows	Yes	Low to moderate	C

PART IV:

projects

the MINDSTORMS method

When creating your own NXT robots, you should use some sort of method, rather than simply diving into a project without a definite sense of direction. A method guides you through the various tasks involved, such as building and programming, to achieve a well-rounded robot. Approaching projects in this manner also avoids needless frustration and confusion.

This chapter introduces what I call the *MINDSTORMS method*, which is a simple strategy I devised for creating not only NXT robots but also MINDSTORMS robots in general. In a sense, this chapter also lays the foundation for the following chapters that present sample projects. Once you understand how to use this method to develop your own NXT robots, you can effectively apply the construction and programming principles demonstrated by the sample projects.

hitting the target with the MINDSTORMS method

The *MINDSTORMS method* consists of four main steps or phases that guide you from the beginning of an NXT project to its completion (Figure 10-1). We'll examine each of these steps and what they involve. While I encourage you to initially adhere to this method as shown, feel free to modify it as you like once you have more experience—there's more than one way to create a robot!

step 1: getting an idea for a robot

You should always begin an NXT project by focusing on an *idea*, one which will serve as the foundation for the project. The essence of a robot is its function—the task it performs—which is why the first question people naturally ask about a robot is, "What does it do?" Therefore, the idea should revolve around the function or purpose of a potential robot.

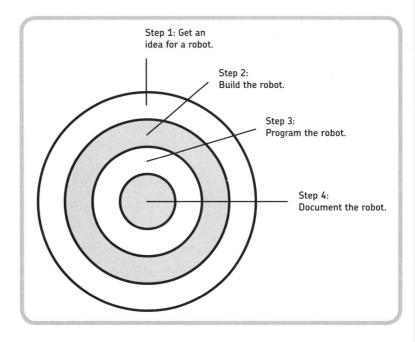

Step 1: Get an idea for a robot.

Step 2: Build the robot.

Step 3: Program the robot.

Step 4: Document the robot.

Figure 10-1: The MINDSTORMS method consists of four main steps.

What if you're not sure what you want to create? Start brainstorming. A great way to gather ideas is by observing the sample projects in the following chapters and from the NXT set (i.e., in the Robo Center on the NXT software). In addition, at the NXTLOG on the MINDSTORMS website, you can browse numerous NXT creations made by other LEGO fans. I'll discuss NXTLOG further in step 4.

NOTE At the most basic level of functionality, NXT robots fall into two broad categories: mobile robots and stationary robots. A *mobile robot*, such as a vehicle, has the ability to move from its starting point to a different location. A *stationary robot*, such as a robotic arm, does *not* have this capability.

step 2: building the robot

Once you have an idea for a robot, the next step is to *build* the robot. Constructing your own robot is largely a process of trial and error in which you experiment with different pieces and building techniques until you achieve the desired results. Along the way you'll be testing what you've built to ensure that it's both durable and functional. It's not uncommon to modify your original idea during this step; once you actually begin building, it sometimes becomes obvious that you'll need to make changes—sometimes significant ones.

In most cases, you should strive to build *modular* robots—robots that consist of subassemblies that you can easily connect and disconnect. Modularity facilitates building

because you can modify sections of your robot without having to dismantle the entire creation. In the following chapters, you'll build a variety of modular robots and see examples of effective subassemblies. For instance, Zippy-Bot—a robot you'll create in Chapter 11—consists of four subassemblies, along with some other pieces mainly used to connect the subassemblies (Figure 10-2).

step 3: programming the robot

Once you've built the robot, the third step is to *program* the robot. To some extent, programming is also a process of trial and error because you're continually testing code to determine what works and what doesn't. With practice, of course, programming becomes easier and you'll make fewer mistakes.

Start by deciding which programming language or languages you'll use, and then begin developing the program. During this phase, you'll be writing, testing, and revising code. Consider the program finished when the robot operates to your satisfaction. At that point, both the program *and* the robot are finished.

It's important to note that building and programming are interrelated to some degree. Sometimes you'll stop building for a moment and do some programming to test what you're building. Other times you'll stop programming for a moment—or longer—and do some building to fix a structural or mechanical issue so that the program will work. In short, don't entirely separate these two activities in your mind.

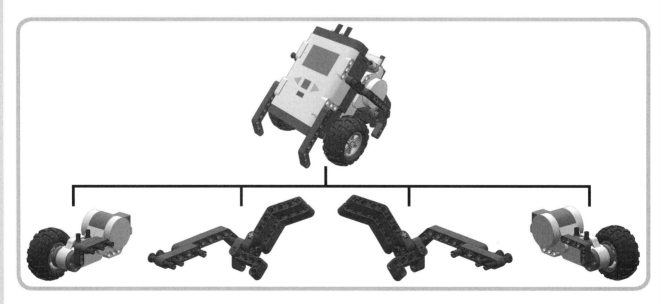

Figure 10-2: Zippy-Bot is composed of four subassemblies as well as some additional pieces that mainly help to connect the subassemblies.

step 4: documenting the robot

The fourth and final step is to *document* the completed robot so that you can not only add it to your portfolio of NXT robots but also share it on the Internet with other LEGO fans. You might start by writing a summary of the robot's function, construction, and programming. If you have a digital camera, snap some pictures of the robot. You could also create building instructions for the robot by taking pictures of each step or using LEGO computer-aided design (CAD) software such as LEGO Digital Designer (LDD), shown in Figure 10-3. Finally, if you have the equipment, consider taking some video of the robot in action.

NOTE The official LDD software (http://ldd.lego.com) and the unofficial LDraw system of tools (http://www.ldraw.org) are two of the most popular LEGO CAD resources. They are offered for free on the Internet, and you can use both of them to create NXT models and building instructions. Consult Appendix C for more LEGO CAD resources.

Documentation doesn't have to be fancy. Some people write lengthy descriptions, take countless pictures, and create detailed building instructions, but you don't have to go to that extent. I encourage you, however, to take at least one picture if you have a camera. When I've failed to do that, I regretted it afterward!

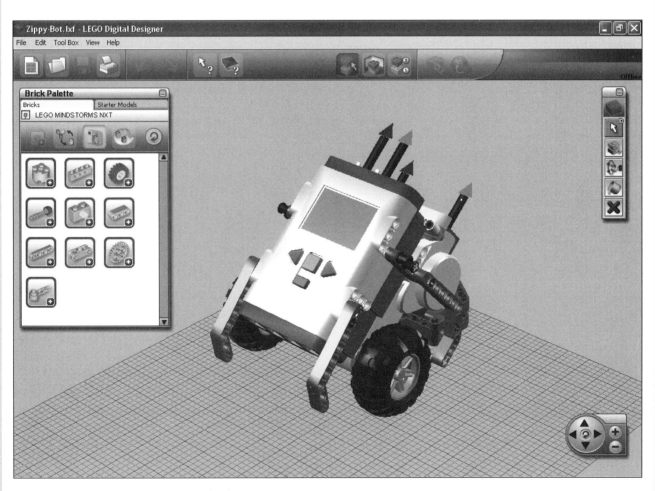

Figure 10-3: Zippy-Bot in LEGO Digital Designer (LDD)

Once you've put together some documentation for the robot, I encourage you to post it on NXTLOG (http://mindstorms.lego.com/nxtlog/default.aspx). NXTLOG allows you to not only submit a description and pictures of the robot, but to upload the robot's program, as well. Often, there are also exciting contests that your NXT creation can win!

conclusion

Creating an NXT robot is a process, and every process needs a good method. You saw in this chapter how to successfully develop NXT robots using the four-step MINDSTORMS method: Get an idea for a robot, build the robot, program the robot, and document the robot. In the next chapter you'll begin the sample projects, which will give you practice working with the NXT set. And practice is essential to becoming a competent NXT inventor!

zippy-bot

Vehicles are one of the most popular types of NXT creations because they're easy to build and fun to watch. Since they are also excellent introductory robots, our first project is a simple vehicle called Zippy-Bot (Figure 11-1). If you like fast robots, this one won't disappoint you! Two motors with attached balloon wheels drive the small robot, and a ball caster (which we'll discuss later) supports the robot in the back. You'll also notice two angled beams on the very front, which prevent the robot from tipping forward. In this chapter, you'll build Zippy-Bot and then program it to perform some basic maneuvers, testing it throughout the programming phase. We'll use this model as the basis for more functional robots in the next two chapters when we add additional pieces, including sensors.

NOTE To give you a clear view of the robots in this book, I only show the ends of the electrical cables rather than the entire cables.

Figure 11-1: Zippy-Bot, a two-wheeled vehicle with a ball caster

building zippy-bot

Zippy-Bot is a modular robot that is composed of four subassemblies. You'll build each of the subassemblies first; then you'll connect them and add some additional pieces in the final assembly. Figure 11-2 is a *Bill of Materials (BOM)* that shows you which pieces are required from the NXT set to construct Zippy-Bot.

These are the four subassemblies:

* Left Drive subassembly
* Right Drive subassembly
* Ball Caster subassembly (part I)
* Ball Caster subassembly (part II)

left drive subassembly

The *Left Drive subassembly* drives the left side of Zippy-Bot; you'll build this subassembly around a servo motor in the following five steps.

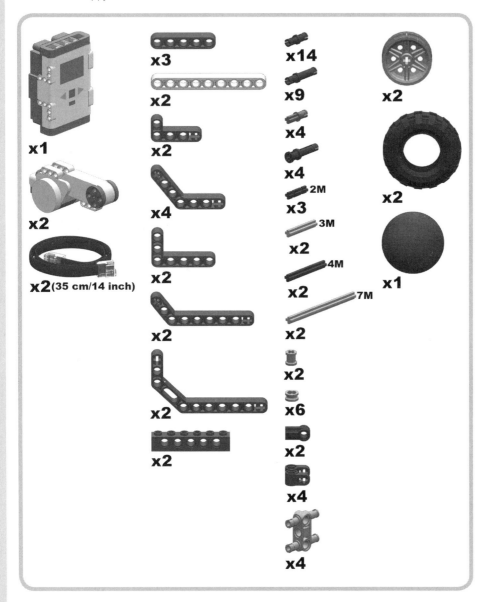

Figure 11-2: A Bill of Materials (BOM) for Zippy-Bot

1

x1 x1

7M

x1 x1

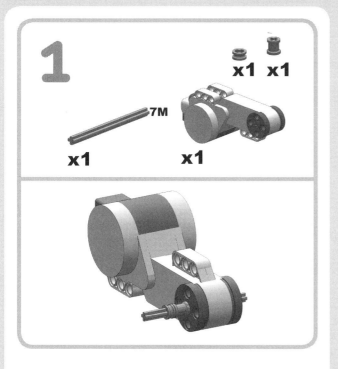

Step 1: Push a 7M axle through a servo motor's output shaft, and then add a bushing and a half-bushing to the axle.

2

x1 x1

Step 2: Add a balloon wheel and a balloon tire to the axle. Make sure that the side of the wheel with spokes is facing away from the motor.

3

x1 x2

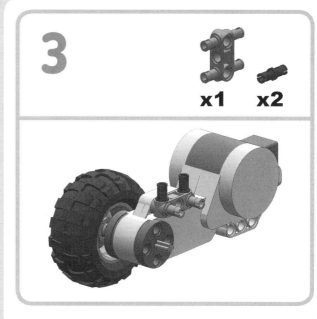

Step 3: Turn the subassembly around, and then add a 3M pegged block and two friction pegs.

4

x1 x3

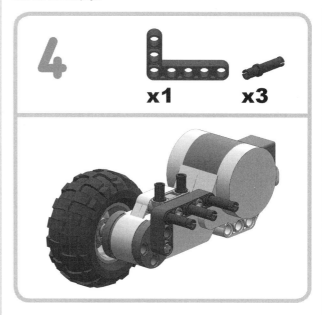

Step 4: Add a 7M perpendicular angled beam, and then snap three 3M friction pegs into the beam.

5

x1

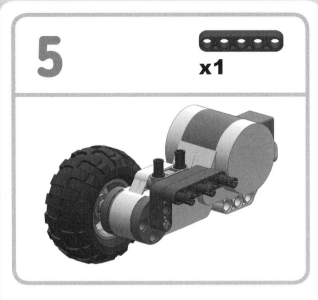

Step 5: Finish the subassembly by attaching a 5M beam to the 3M friction pegs.

right drive subassembly

As you might have guessed, the *Right Drive subassembly* drives the right side of Zippy-Bot. There are four steps for building this subassembly.

1

7M

x1   x1

x1   x1

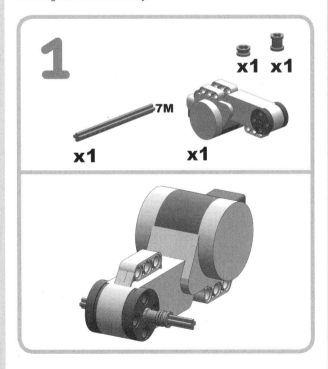

Step 1: Push a 7M axle through a servo motor's output shaft, and then add a bushing and a half-bushing to the axle.

2

x1   x1

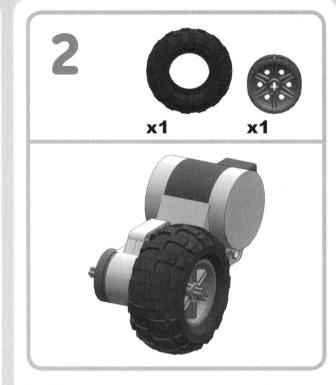

Step 2: Add a balloon wheel and a balloon tire to the axle. Make sure that the side of the wheel with spokes is facing away from the motor.

3

x1   x2

Step 3: Turn the subassembly around, and then add a 3M pegged block and two friction pegs.

4

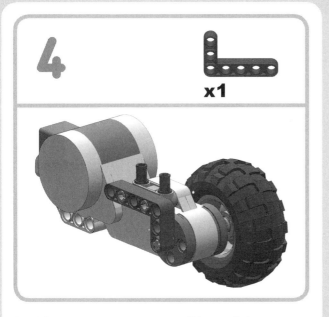

Step 4: Finish the subassembly by adding a 7M perpendicular angled beam.

ball caster subassembly (part I)

The *Ball Caster subassembly* is composed of two parts that we'll build separately and then connect in the final assembly. This subassembly holds one of the plastic balls from the NXT set directly behind Zippy-Bot, and the robot leans backward on the ball. Then when Zippy-Bot moves, the ball rotates like a third wheel. Unlike a regular *caster* or swiveling wheel (such as the one TriBot uses), a *ball caster* remains stationary while the ball itself can freely spin in any direction. Follow these four steps to build the first part of the Ball Caster subassembly.

1

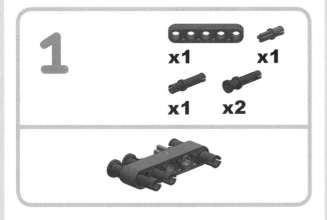

Step 1: Take a 5M beam and add two bushed friction pegs, a friction axle peg, and a 3M friction peg.

2

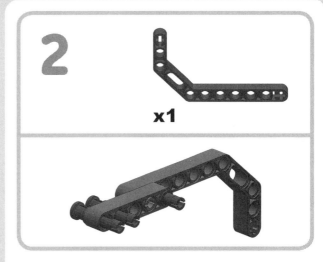

Step 2: Snap on an 11.5M angled beam.

3

2M

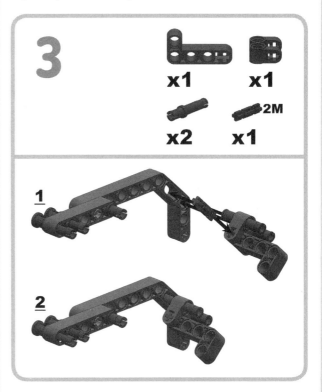

Step 3: Put two 3M friction pegs into a split cross block, push a 2M notched axle through the split cross block and a 5M perpendicular angled beam, and then snap all of it into the 11.5M angled beam.

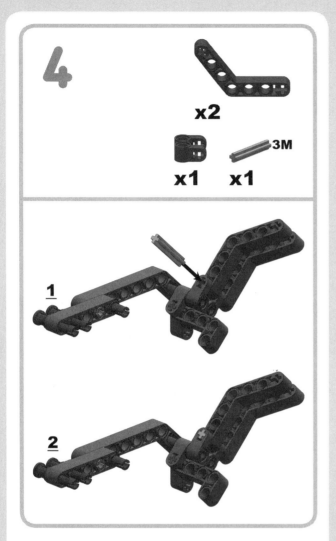

Step 4: Snap another split cross block onto the 3M friction pegs, and then push a 3M axle through the split cross block and two 7M angled beams.

ball caster subassembly (part II)

The second part of the Ball Caster subassembly *mirrors* the first part, which means it uses all the same pieces but the orientation of the pieces is switched. In the final assembly, each part will attach to a different side of the robot. Follow these four steps to build the second part of the Ball Caster subassembly.

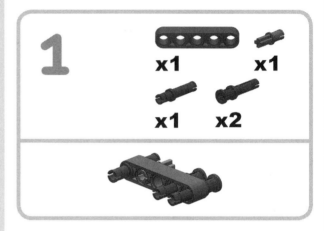

Step 1: Take a 5M beam and add two bushed friction pegs, a friction axle peg, and a 3M friction peg.

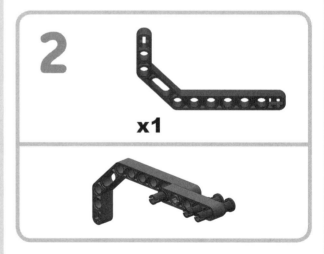

Step 2: Snap on an 11.5M angled beam.

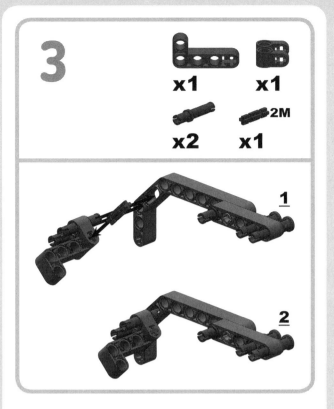

3

x1 x1

x2 x1 2M

1

2

Step 3: Put two 3M friction pegs into a split cross block, push a 2M notched axle through the split cross block and a 5M perpendicular angled beam, and then snap all of it into the 11.5M angled beam.

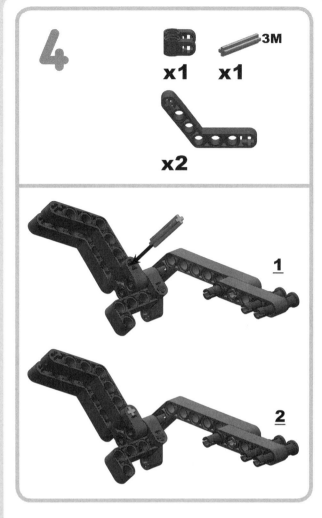

4

x1 x1 3M

x2

1

2

Step 4: Snap another split cross block onto the 3M friction pegs, and push a 3M axle through the split cross block and two 7M angled beams.

final assembly

Now it's time to complete the robot by connecting the subassemblies and adding some additional pieces. Steps 1 through 11 show you how to do it.

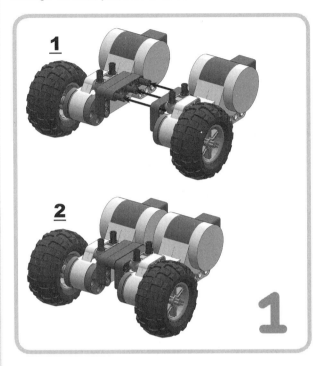

Step 1: Connect the Right Drive subassembly to the Left Drive subassembly.

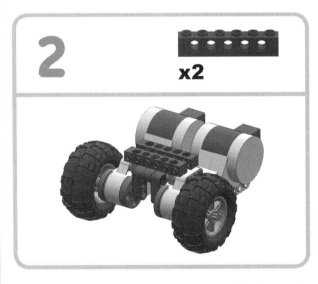

Step 2: Use two 1 × 6 TECHNIC bricks to further strengthen the connection between the drive subassemblies.

Step 3: Flip the model over, and then add two 3M pegged blocks and four friction pegs to the backs of the motors.

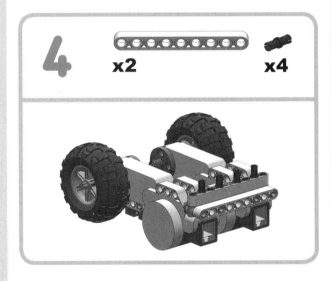

Step 4: Add two 9M beams and four friction pegs.

5

x1

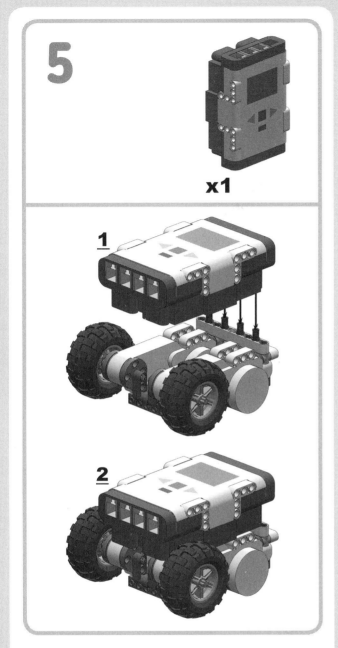

Step 5: Turn the model around, and then push the NXT into the friction pegs using the round-holes on the NXT's bottom side.

6

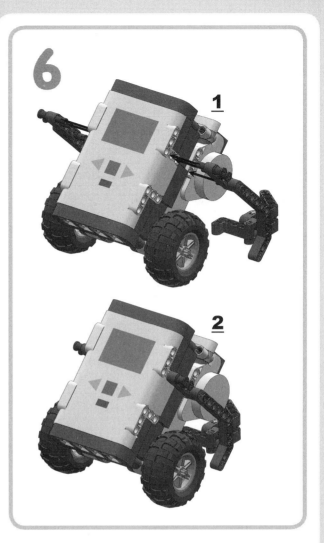

Step 6: Attach both parts of the Ball Caster subassembly to the robot. Each one snaps into a motor and one side of the NXT. In this way, the NXT is even more firmly connected to the rest of the robot.

7

x2 x2

Step 7: Add two friction pegs and two friction axle pegs to the front of the NXT.

8

x2

Step 8: Add two 9M angled beams to the front of the NXT.

9

4M

x2 x2

Step 9: Turn Zippy-Bot around, and then add two 4M axles and two half-bushings to the end of the Ball Caster subassembly. You'll need to add one axle at a time, positioning a half-bushing and then pushing the axle through it.

10

x2 x2 x1 2M

Step 10: Connect two #1 angle connectors with a 2M notched axle, and then put them on top of the axles from step 9. Attach two more half-bushings to the axles.

11

x2 (35 cm/14 inch) x1

Step 11: Put a ball from the NXT set underneath the Ball Caster sub-assembly, and then connect the motors to the appropriate output ports (C and B) with 35 cm/14 inch cables.

programming and testing zippy-bot

Currently, Zippy-Bot is limited in what it can do because it has no sensors and only two motors. But rather than focusing on what Zippy-Bot *cannot* do, let's focus on what it can do—move! In the following three NXT-G programs, we'll practice using the servo motors to precisely control Zippy-Bot.

NOTE You can download all the programs for Zippy-Bot and the other robots in this book from this book's companion website (http://www.nxtguide.davidjperdue.com).

driving in a straight line

As you learned in Part III, the servo motors enable mobile robots like Zippy-Bot to easily travel in a straight line. We'll use Zippy-Bot with the test pad from the NXT set to try out this highly useful capability. Although the test pad has a number of features, right now we're interested in the starting position at the front and the picture of a red ball near the other side. The goal will be for Zippy-Bot to drive in a straight line from the starting point to the red ball and then back to the starting point. To see how precisely the servo motors performed, just compare where Zippy-Bot ended to where Zippy-Bot began.

We'll program Zippy-Bot with the Move block, which can synchronize the two motors. One Move block will drive Zippy-Bot forward 3 rotations, and then the next Move block will drive Zippy-Bot backward 3 rotations.

NOTE Because of Zippy-Bot's design, the forward setting on the Move block drives Zippy-Bot backward, while the reverse setting drives Zippy-Bot forward. As you observe the Zippy-Bot1 program in Figure 11-3, remember that the arrows on the left side of each block indicate the direction of rotation for the motors, not necessarily the direction in which the robot will be moving.

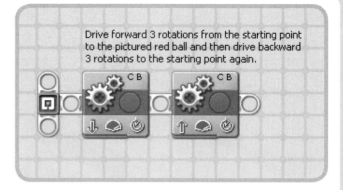

Figure 11-3: The Zippy-Bot1 program drives Zippy-Bot forward and then backward.

Download the Zippy-Bot1 program to Zippy-Bot and get out your test pad. At the front of the test pad is the starting point with the word *START* in large letters. Position Zippy-Bot so that it faces the red ball at the other end of the test pad with its ball caster directly over the word *START*. Try to make sure that Zippy-Bot is as straight as possible.

NOTE Although the ball used by Zippy-Bot for its ball caster does stay in place while the robot operates, it isn't attached to its subassembly, so it will fall out when you pick up the robot.

Now run the program. Zippy-Bot should shoot forward, stop at the red ball, and then dash backward, coming to a complete stop at the starting point. How close is Zippy-Bot to its exact starting position? If you set up the robot properly, it should be very close. Run the program a few more times to verify the results.

changing direction by spinning in place

Instead of simply driving backward after reaching the red ball, Zippy-Bot could also turn around and then drive forward to the starting point. After all, mobile robots don't always move in straight lines! Figure 11-4 shows the Zippy-Bot2 program, which has six programming blocks. After driving forward 3 rotations and playing a tone, Zippy-Bot spins in place for 0.88 rotations by rotating its wheels in

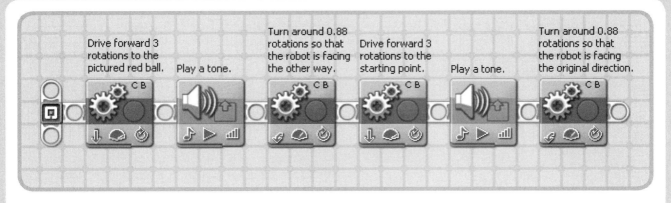

Drive forward 3 rotations to the pictured red ball.

Play a tone.

Turn around 0.88 rotations so that the robot is facing the other way.

Drive forward 3 rotations to the starting point.

Play a tone.

Turn around 0.88 rotations so that the robot is facing the original direction.

Figure 11-4: The Zippy-Bot2 program demonstrates how to make Zippy-Bot turn by spinning in place.

opposite directions. (Moving the Move block's Steering parameter slider to the leftmost or rightmost position causes the wheels to rotate in opposite directions.) Then Zippy-Bot drives forward 3 rotations to the starting point, plays a tone, and turns around again so that it faces the original direction.

NOTE Through experimentation, I found that setting the Duration parameter to 0.88 rotations works well for turning Zippy-Bot around so that it faces the opposite direction. Likewise, when programming your own robots, you'll often need to determine durations and other settings by trial and error.

Download the Zippy-Bot2 program to Zippy-Bot and position the robot as you did before: Zippy-Bot should face the red ball on the test pad, and its ball caster should be directly over the word *START*. Run the program and watch what happens. After Zippy-Bot has driven to the red ball and turned around, how accurately does it return to the

starting position? Zippy-Bot should successfully return to the general starting point area. Try repositioning Zippy-Bot as before and running the program again. Are the results the same? You'll find that the servo motors are very precise but not *perfectly* precise due to a variety of factors (such as slippage). Nevertheless, they are still immensely useful.

changing direction by steering

Instead of spinning in place to change its direction, Zippy-Bot can also steer left or right as it drives. If both wheels are rotating in the same direction but one is spinning faster than the other, Zippy-Bot steers toward the side of the slower wheel. The faster one wheel rotates relative to the other wheel, the more sharply Zippy-Bot turns. Moving the Move block's Steering parameter slider to the left or right (but *not* to the leftmost or rightmost positions) steers Zippy-Bot in this manner. In the Zippy-Bot3 program (Figure 11-5), which has five programming blocks, two different Move blocks steer Zippy-Bot.

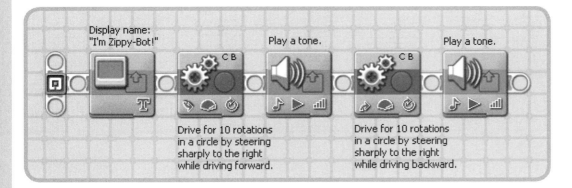

Display name: "I'm Zippy-Bot!"

Play a tone.

Play a tone.

Drive for 10 rotations in a circle by steering sharply to the right while driving forward.

Drive for 10 rotations in a circle by steering sharply to the right while driving backward.

Figure 11-5: The Zippy-Bot3 program drives Zippy-Bot in circles by steering.

The Display block at the beginning of the program displays *I'm Zippy-Bot!* on the NXT's LCD (even robots need some personality). Zippy-Bot then drives for 10 rotations in a circle by steering sharply to the right while moving forward. After playing a tone, Zippy-Bot essentially retraces its steps by driving for 10 rotations in a circle but moving backward. The last block is another Sound block that plays a tone, letting you know that the program has finished.

Download the Zippy-Bot3 program to Zippy-Bot and place the robot on the floor (both carpet and hard surfaces work fine), making sure Zippy-Bot has some room to move around. Run the program and watch as Zippy-Bot drives in a circle by steering; notice how the wheels rotate at different speeds. When the program finishes, feel free to experiment with the Move blocks' settings to make Zippy-Bot turn in other ways.

conclusion

Zippy-Bot provides an example of a very simple yet functional NXT vehicle. In this chapter, you built Zippy-Bot by constructing the Left Drive subassembly, the Right Drive subassembly, and the Ball Caster subassembly (in two different parts) and then connecting them in the final assembly with additional pieces. You also programmed Zippy-Bot to perform a variety of maneuvers and learned how to use the servo motors to achieve precision. In the next chapter, we'll develop Zippy-Bot into a unique robot with a specific function by adding more pieces and writing more code.

12

bumper-bot

Given how effectively they move, NXT vehicles can competently perform a multitude of tasks: transporting objects, following a line, traversing a maze, and so much more. Among the many potential projects for this subcategory of robots, one in particular is perhaps the classic MINDSTORMS project: a vehicle that explores a room and detects obstacles with a bumper that is built around a single touch sensor. In this chapter, you'll create Bumper-Bot (Figure 12-1) by adding a bumper to Zippy-Bot and then programming the robot to explore its surroundings.

Figure 12-1: Bumper-Bot explores a room and detects obstacles with a bumper on its front end.

building bumper-bot

To build Bumper-Bot, you need the basic Zippy-Bot model from Chapter 11 and one additional subassembly: the Bumper subassembly. After you've built this subassembly, you'll combine it with the Zippy-Bot model to create Bumper-Bot. Figure 12-2 is a BOM for Bumper-Bot, showing all the pieces required for this project.

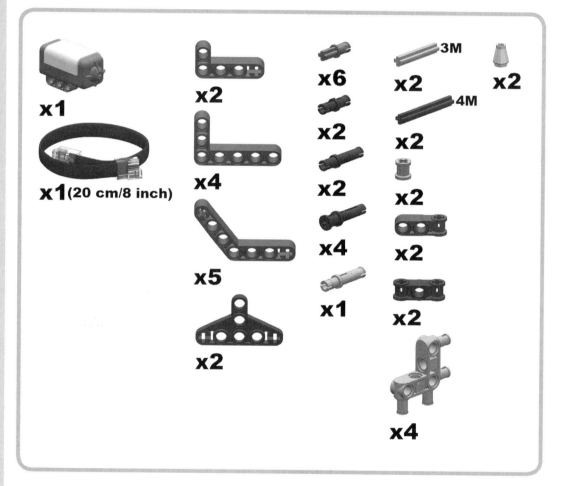

Figure 12-2: A BOM for Bumper-Bot

bumper subassembly

When Bumper-Bot runs into an object, the *Bumper subassembly* transfers the impact to the touch sensor, pressing its push button. The robot immediately realizes, "Ah! I've run into something," and it will then take the appropriate action. Some of the most effective bumper designs use two (or more) touch sensors, but the NXT set comes packaged with only one touch sensor. Nevertheless, we can still create an effective bumper by extending pieces sideways to increase the range of detection. There are 10 steps to build this subassembly.

Step 1: Add a 7M angled beam with a 3M peg, a 3M friction peg, and a 3M axle to the touch sensor.

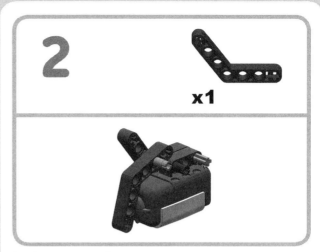

Step 2: Attach another 7M angled beam to the 3M peg. This beam, which can swivel freely on the 3M peg, is the piece that actually presses the touch sensor's push button.

Step 3: Add another 7M angled beam, and then push two 5M pegged perpendicular blocks into the 7M angled beams.

4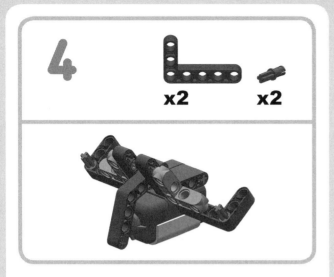

x2 x2

Step 4: Add 7M perpendicular angled beams to the pegged blocks, and then add friction axle pegs to the 7M perpendicular angled beams.

5

x2 x2 x2 4M

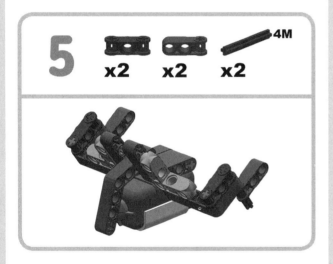

Step 5: Add two inverted cross blocks, push two 4M axles through them, and then connect two extended cross blocks to the axles.

6

x4 x2

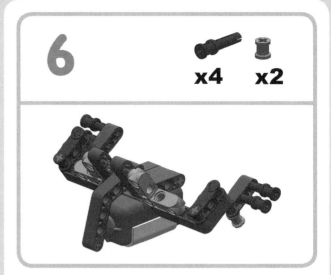

Step 6: Add four bushed friction pegs to the extended cross blocks and two bushings to the ends of the 4M axles. You'll use the bushed friction pegs later to easily add this subassembly to Zippy-Bot.

7

x1 x1 3M

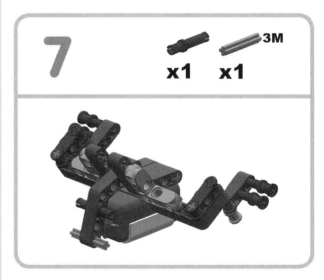

Step 7: At the front of the model, add a 3M friction peg and a 3M axle to the 7M angled beam.

8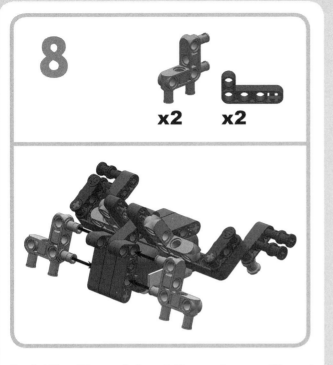

x2   **x2**

Step 8: Add two 5M perpendicular angled beams and two more 5M pegged perpendicular blocks.

10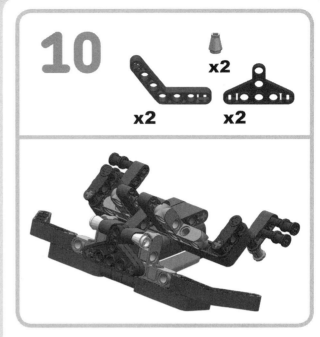

x2

x2   **x2**

Step 10: Finish by adding two 7M angled beams, two triangular half-beams, and two 1 × 1 cones (these will be the robot's "eyes"). You'll need to push the cones into the round-holes.

9

x2   **x2**   **x4**

Step 9: Add two more 7M perpendicular angled beams, two friction pegs, and four friction axle pegs.

final assembly

It's easy to attach Zippy-Bot to the Bumper subassembly, so the final assembly consists of only two steps.

x1 (20 cm/8 inch)

Step 1: Use a 20 cm/8 inch electrical cable to connect the touch sensor on the Bumper subassembly to the NXT (input port 1).

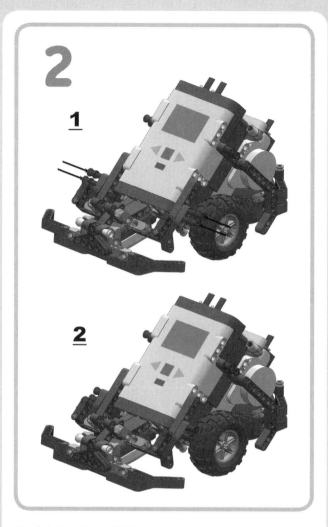

Step 2: Push the bushed friction pegs on the Bumper subassembly into the angled beams on the front of Zippy-Bot. You're done!

NOTE As you attach the Bumper subassembly to Zippy-Bot, make sure that you keep the excess length of the touch sensor's electrical cable in front of the robot. If you push the extra length of the cable behind the subassembly, the robot's right wheel could get caught on it.

programming and testing bumper-bot

Although I said at the beginning of the chapter that Bumper-Bot explores its surroundings, the robot cannot determine its location or current direction. For Bumper-Bot, exploration means driving forward and changing direction whenever it hits an obstacle with its bumper. We'll examine two NXT-G programs that present variations on this approach.

exploring: the basic approach

Our first program, Bumper-Bot1 (Figure 12-3), begins by waiting for you to press and release the bumper. When you do, the NXT plays a tone and the main part of the program gets under way. In a Loop block that repeats forever, there are five blocks that control the robot's movement. First, a Move block drives Bumper-Bot forward until the robot runs into an obstacle with its bumper, at which point the NXT plays the Sonar sound. Bumper-Bot immediately drives backward for 1 rotation and then spins in place for

0.5 rotations. At that point, of course, the Loop block begins the series of blocks all over again.

NOTE Remember that movement measurements, such as 0.5 rotations or 180 degrees, refer to the rotation of the motors (i.e., the motors' output shafts), not the robot as a whole.

Download the Bumper-Bot1 program to Bumper-Bot, and place the robot on the floor, making sure it has some room to move around. Run the program, press and release the bumper, step back, and watch what happens. Bumper-Bot should easily detect walls and box-shaped objects, but some pieces of furniture such as chairs could be problematic. If Bumper-Bot gets stuck, untangle it as soon as possible to prevent the motors from being in a stall situation too long. When you're finished testing the program, you'll need to stop it by pressing the Clear/Go Back button on the NXT.

exploring: a more advanced approach

While BumperBot1 certainly works, it can be somewhat predictable by causing Bumper-Bot to travel in the same area or pattern. With the Random block from the Complete palette,

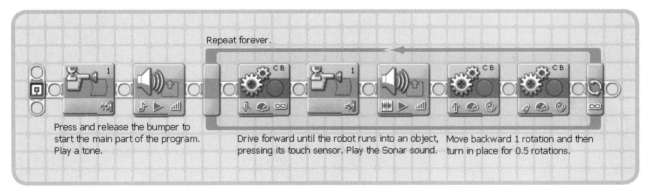

Figure 12-3: Bumper-Bot1 drives Bumper-Bot in a straight line until the robot hits an object with its bumper, which causes Bumper-Bot to drive backward a small distance, turn around, and then continue driving forward.

we can introduce some random movement. Toward the end of the improved program, Bumper-Bot2 (Figure 12-4), there is a Random block that sends its output via a data wire to the Duration plug on the following Move block. This Random block sends a value ranging from 120 to 480 to the Move block, which uses this value to determine how many degrees the robot should turn in place.

Download the Bumper-Bot2 program to Bumper-Bot, and place the robot on the floor again. Run the program, press and release the bumper, step back, and watch Bumper-Bot perform. (As before, be ready to untangle Bumper-Bot when necessary.) After hitting an object and backing up, Bumper-Bot should now spin in place for different durations, making its movement much less predictable. Finally, when you're finished testing the program, press the Clear/Go back button on the NXT to stop it.

conclusion

In this chapter, you built and programmed Bumper-Bot, a vehicle equipped with a bumper that uses a touch sensor to detect obstacles. After constructing the Bumper subassembly and adding it to Zippy-Bot, you programmed the robot to explore its surroundings, first without random movement and then with random movement. In the next chapter, you'll create another robot based on Zippy-Bot—this time using two sensors.

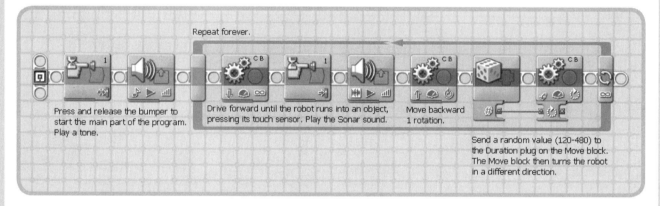

Figure 12-4: Bumper-Bot2 adds a Random block to achieve random movement.

13

claw-bot

Claw-Bot is a hunting vehicle that is also based on Zippy-Bot (Figure 13-1). Dwelling on the inside of the NXT test pad (i.e., the area on the test pad encompassed by the black line), Claw-Bot hunts for objects, and when it finds them, Claw-Bot pushes them out of its abode. Claw-Bot uses an ultrasonic sensor to detect objects and a light sensor to detect its boundary, the black line. In addition, the claw-like structure on the front helps the robot to push objects around. We'll start this chapter by constructing Claw-Bot, and then we'll proceed with programming and testing.

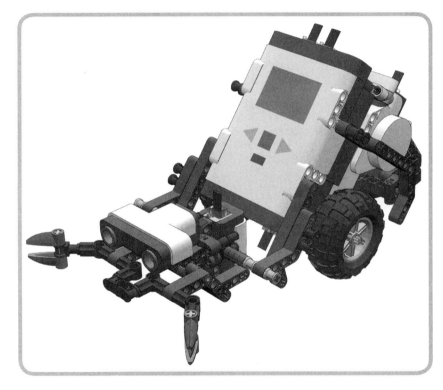

Figure 13-1: Claw-Bot is a hunting vehicle that detects objects with the ultrasonic sensor.

building
claw-bot

In addition to Zippy-Bot from Chapter 11, you'll need the pieces shown in Figure 13-2 to build Claw-Bot. There is only one additional subassembly that you need to construct, the Claw subassembly. You'll attach this subassembly to the Zippy-Bot model to create Claw-Bot.

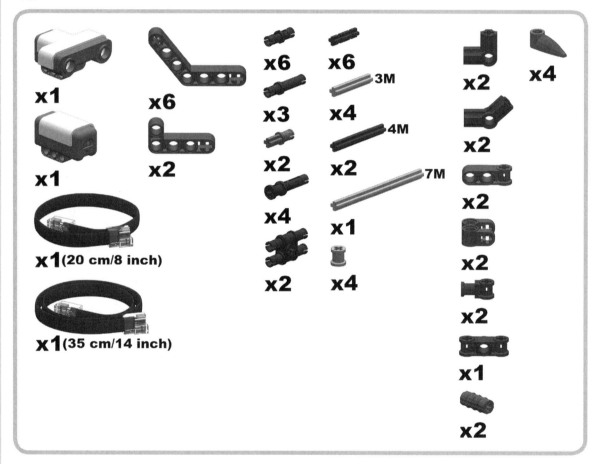

Figure 13-2: A BOM for Claw-Bot

claw subassembly

The *Claw subassembly* is relatively compact and features the ultrasonic sensor, the light sensor, and the characteristic "claws." There are 14 steps for this subassembly.

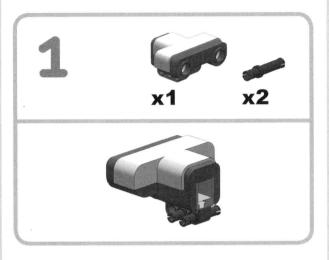

Step 1: Push two 3M friction pegs into the back of the ultrasonic sensor.

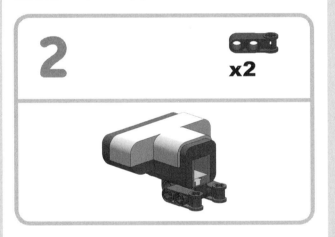

Step 2: Add two extended cross blocks.

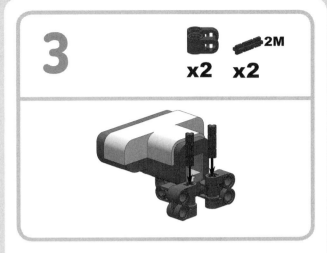

Step 3: Add two split cross blocks and two 2M notched axles. You must secure one split cross block at a time, first positioning it and then pushing an axle through the cross-holes.

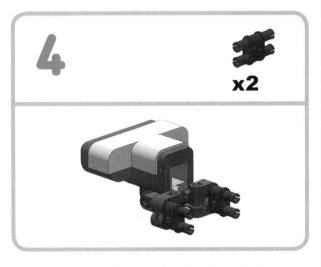

Step 4: Snap two double friction pegs into the split cross blocks.

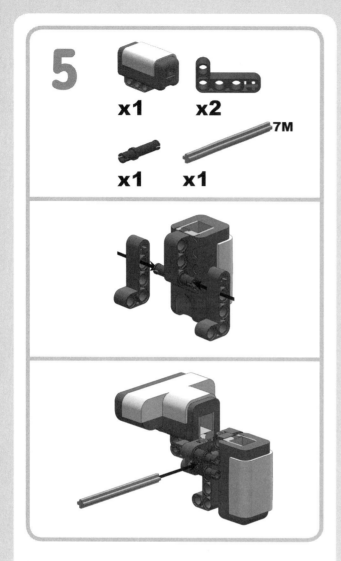

Step 5: Attach a 3M friction peg and two 5M perpendicular angled beams to the light sensor. Then align the cross-holes on the 5M perpendicular angled beams with the cross-holes on the double friction pegs, and push a 7M axle through them.

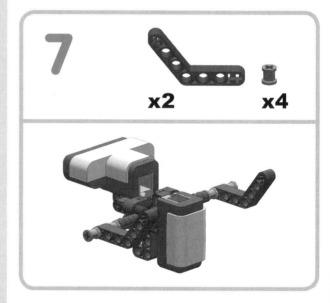

Step 6: Put two 4M axles in two axle extenders, and then push the axle extenders onto the 7M axle.

Step 7: Attach two 7M angled beams and four bushings to the 4M axles. You're halfway done!

8

x4

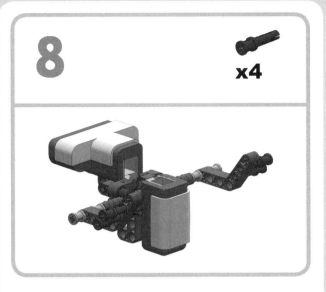

Step 8: Push four bushed friction pegs into the angled beams, but don't insert them all the way. You'll push them through completely during the final assembly when you attach the Claw subassembly to Zippy-Bot.

9

x4

Step 9: Turn the model around to the front. Add four friction pegs to the 5M perpendicular angled beams.

10

x2 x2 x2

Step 10: Add two 7M angled beams with two friction pegs and two friction axle pegs.

11

3M

x2 x2

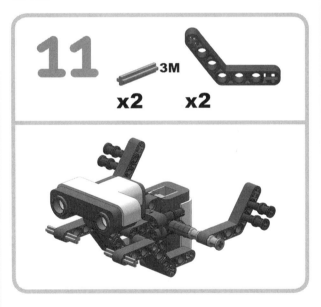

Step 11: Add two more 7M angled beams and two 3M axles.

12 x2 x2 x4 2M

Step 12: Add two #4 and two #6 angle connectors, along with four 2M notched axles.

13 x1 x2 x2 3M

Step 13: Add an inverted cross block, two catches, and two 3M axles.

14 x4

Step 14: Finish the Claw subassembly by pushing four TECHNIC teeth onto the 3M axles.

final assembly

It's easy to attach the Claw subassembly to Zippy-Bot, so the final assembly consists of only three steps. Make sure, however, that you carefully follow the instructions regarding the electrical cables.

1

x**1**(20 cm/8 inch)

x**1**(35 cm/14 inch)

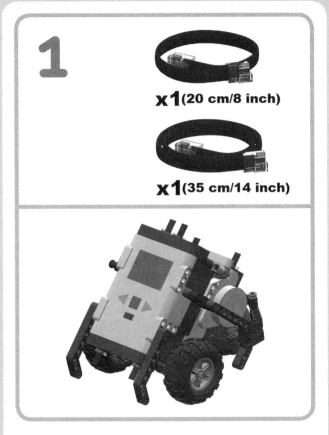

Step 1: Take Zippy-Bot and connect a 20 cm/8 inch electrical cable to input port 1 and a 35 cm/14 inch electrical cable to input port 4. The cables will hang freely until you connect their other ends to the sensors in step 3.

2

Step 2: Push the bushed friction pegs on the Claw subassembly into the angled beams on the front of Zippy-Bot.

3

Step 3: Connect the electrical cables from step 1 to the light sensor (input port 1) and ultrasonic sensor (input port 4). Claw-Bot is finished and ready to do some hunting!

NOTE Once you've attached an electrical cable to the ultrasonic sensor, the only way to unplug it is by pushing an axle (8M or longer works well) from underneath the subassembly to press and release the cable's prong.

programming and testing claw-bot

With construction finished, you're prepared to program and test Claw-Bot. The first program we'll examine employs a basic approach to finding objects and pushing them out of the area encompassed by the black line. Then we'll observe a slightly more advanced program that provides greater effectiveness. For the sake of simplicity, we'll assume that Claw-Bot always begins at the starting point on the test pad. In addition, we'll assume that objects could be anywhere within the boundary of the black line and that Claw-Bot must do its hunting within that boundary. When pushing objects out of bounds, however, Claw-Bot can go beyond the black line.

NOTE Remember that you can download all the programs for the robots in this book from this book's companion website (http://www.nxtguide .davidjperdue.com).

hunting and pushing objects: the basic approach

Our first program, Claw-Bot1, instructs Claw-Bot to hunt for three objects and push each one out of bounds (Figure 13-3). Claw-Bot1 begins with a Move block that drives Claw-Bot to the center of the test pad. Next, within a Loop block that repeats three times (once for each object), a series of programming blocks tells the robot to do the following:

1. Spin in place until the ultrasonic sensor detects an object less than 11 inches away.

2. Travel forward, pushing the object, until the light sensor detects the line.

3. Continue traveling forward 0.75 rotations to push the object well out of sight.

4. Return to the center of the test pad, driving backward 2 rotations.

Importantly, the program must have a threshold value for determining when the robot has crossed the black line. For example, if the light sensor reads 38 percent on the black line, an effective threshold value would be 38 + 2, or 40 percent (adding 2 makes Claw-Bot more sensitive to the line). Whenever Claw-Bot's light sensor reads values less than 40 percent, the robot knows it has reached the line. I did, in fact, choose 40 percent for the threshold, but you may need to adjust this value to compensate for lighting differences. You'll learn more about this in a moment.

Download the Claw-Bot1 program to Claw-Bot and position the robot at the starting point on the test pad, so that it faces straight ahead and has its ball caster directly over the word *START*. Now find three relatively small and lightweight objects and put them in various places inside the boundary of the black line. As we discussed back in Chapter 3 (see "Digital Sensor" on page 27), the ultrasonic sensor can generally detect square or box-like objects most effectively. Do *not* try to use books, however, because they can be more difficult for Claw-Bot to detect and push.

Run the program, and watch as Claw-Bot hunts for the objects and pushes them out of its territory (stand back!). Did it successfully find all three objects? If so, did it successfully push all three objects out of bounds? Try running the program several times, setting up the objects in different positions each time.

NOTE Sometimes Claw-Bot mistakes the red ball on the test pad for the black line. This is more likely to happen if the room in which you are testing is dimly lit.

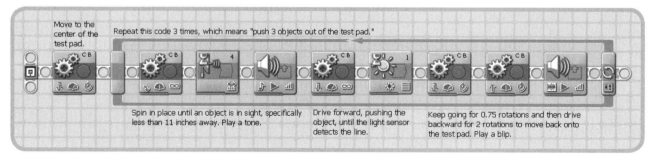

Figure 13-3: The relatively simple Claw-Bot1 program instructs Claw-Bot to hunt for three different objects and push them out of its "territory," the area of the test pad encompassed by the black line.

If Claw-Bot seems to have trouble detecting the black line, the threshold value is probably too low. If Claw-Bot stops before reaching the line (i.e., it thinks it sees the line but it really doesn't), the threshold is probably too high. To adjust the threshold, you first need to take a reading of the line with the light sensor. Selecting the View menu on the NXT, then the Reflected light option, and finally the Port 1 option displays the light sensor's reading as a percentage on the LCD. Next, in the Claw-Bot1 program, click the Wait block configured for a light sensor, and update the value in the Until parameter (remember to add 2 to the reading as we discussed earlier). You may need to adjust the threshold several times to achieve satisfactory results.

hunting and pushing objects: a more advanced approach

Although Claw-Bot1 does work, Claw-Bot occasionally detects an object and then "misses" it while driving forward to the black line. In other words, sometimes Claw-Bot doesn't push anything! Figure 13-4 shows the improved Claw-Bot2 program, which uses a Switch block to help solve this problem. The improved technique for hunting objects and pushing them out of bounds is as follows:

1. Spin in place until the ultrasonic sensor detects an object less than 11 inches away.

2. Travel forward for 0.75 rotations.

3. If the ultrasonic sensor still sees the object, continue to move forward, pushing the object until the light sensor detects the boundary. If the ultrasonic sensor *doesn't* see the object, spin in place until the object is in sight again and then drive forward, pushing the object until the light sensor detects the boundary.

4. Continue traveling forward 0.75 rotations to push the object well out of sight.

5. Return to the center of the test pad by driving backward for 2 rotations.

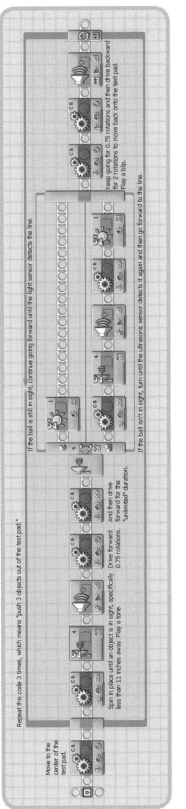

Figure 13-4: The improved Claw-Bot2 program helps Claw-Bot to more effectively push objects out of bounds.

Download the Claw-Bot2 program to Claw-Bot and position the robot at the starting point on the test pad as before. Place the same three objects you used before anywhere inside the boundary of the black line. Run the program, step back, and watch what happens. Claw-Bot should now perform better because it corrects its direction when necessary. You may need to run the program several times, setting up Claw-Bot and the objects each time, before Claw-Bot has to correct its direction. It's important to note that Claw-Bot is more likely to correct its direction when the objects are farther apart.

Finally, it's worth mentioning that you can easily change the number of objects Claw-Bot searches for by selecting the Loop block in the program and changing the number of times the block should repeat. For example, typing 8 in the Loop block's Until parameter tells Claw-Bot to hunt for eight objects instead of three.

conclusion

Claw-Bot demonstrates one more of the many possible tasks that an NXT vehicle can perform. In this chapter, you built the Claw subassembly and attached it to Zippy-Bot to create the complete Claw-Bot. You then programmed Claw-Bot—using a basic program and then a more advanced program—to hunt for three objects on the test pad and push them out of the area encompassed by the black line. In the next chapter, you'll begin building new types of robots that are not based on Zippy-Bot.

tag-bot

Most NXT robots interact with their environments using the variety of available sensors, and designing a robot that interacts with humans is especially fun. You'll create another mobile robot in this chapter; this time it's a robot named Tag-Bot (Figure 14-1) that plays flashlight tag. In this game, your goal is to shine a flashlight on Tag-Bot's light sensor for a short period of time, which "tags" the robot. Tag-Bot's goal is to dash around the room in an attempt to escape the light. When tagged, the robot says, "Game over" and then stops (don't worry—it won't chase you!).

Tag-Bot employs a *steering drive* in which the front wheels steer and the back two driven wheels are fixed. This configuration provides great stability, but it means that the robot cannot make tight turns or turn in place. Nevertheless, a steering drive is a viable solution for many types of projects. Positioned on the front of Tag-Bot are the light sensor, a sound sensor for activating the robot (e.g., by clapping your hands), and a rotating ultrasonic sensor for detecting objects. We'll begin this chapter by building Tag-Bot, and then we'll program and test the robot to perform basic and more advanced behaviors.

Figure 14-1: Tag-Bot plays a game of flashlight tag.

building tag-bot

Tag-Bot is composed of eight subassemblies and uses all three servo motors in the NXT set: Two motors are for driving and one is for steering. Figure 14-2 shows the BOM for Tag-Bot. Ideally, Tag-Bot would use a differential gear for the back wheels. During a turn, the robot's wheels travel different distances (the wheel on the outside travels the greater distance), and without a differential to adjust the distribution of power to the wheels, one wheel must slip. There isn't a differential gear in the NXT set, so Tag-Bot simply does without. While this approach may not provide maximum efficiency, it does work!

NOTE It's possible to build a "homemade" differential with pieces in the NXT set. I chose not to use one for Tag-Bot, however, because of the complexity it would have introduced. If you'd like to tackle that project, you can find building instructions at http://nxtasy.org/2006/08/15/differential.

Now you're ready to build the subassemblies and then combine them in the final assembly. You'll construct the following eight subassemblies in the order shown:

* Left Drive subassembly
* Right Drive subassembly
* Steering subassembly
* Steering Motor subassembly
* Frame subassembly
* Ultrasonic Sensor subassembly
* Light Sensor subassembly
* Sound Sensor subassembly

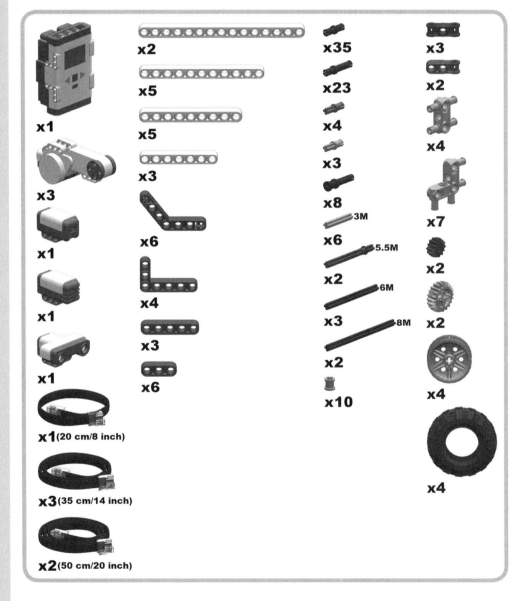

Figure 14-2: A BOM for Tag-Bot

left drive subassembly

The *Left Drive subassembly* drives the left side of the robot. It has a gear train consisting of two gears—a 20t double bevel gear and a 12t double bevel gear—that gear up, resulting in a 3:5 gear ratio (see "Controlling a Gear Train's Performance" on page 54 for a discussion of gearing up and down). Does this ratio provide enough torque to effectively drive the robot? Given the strength of the servo motors and the fact that we're using two of them to drive the robot (one is included in the next subassembly), this gear ratio does in fact provide sufficient torque. Complete the following seven steps to build this subassembly.

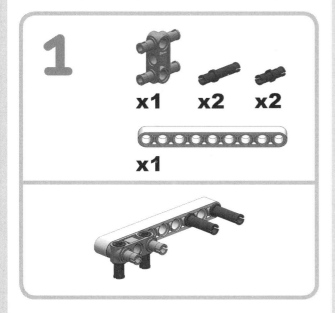

Step 1: Begin by snapping a 3M pegged block and two 3M friction pegs into a 9M beam. Next, add two friction pegs to the pegged block.

Step 2: Add a 7M angled beam, two 3M friction pegs, and a 3M axle.

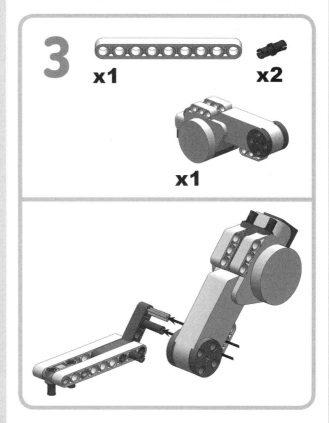

Step 3: Add another 9M beam and a motor to the 7M angled beam. Then connect two friction pegs to the top of the motor.

4

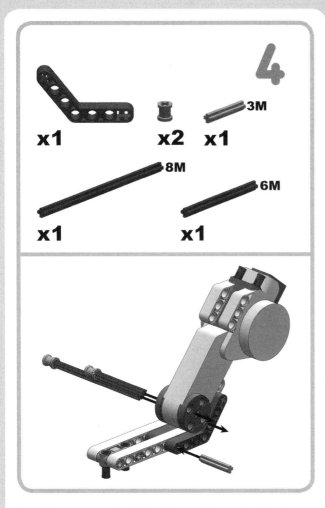

x1   x2   x1 **3M**

8M
x1

6M
x1

Step 4: First, attach another 7M angled beam. Next, push a 3M axle through the two 7M angled beams. Finally, add bushings to 8M and 6M axles and then push them through as shown, with the 6M axle running through the motor's output shaft and the 8M axle running through the 7M angled beams.

5

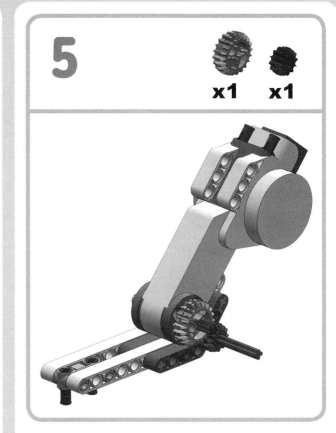

x1   x1

Step 5: Connect a 20t double bevel gear to the 6M axle and a 12t double bevel gear to the 8M axle.

6

x1   x1

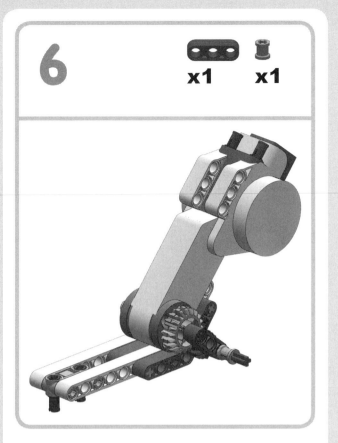

Step 6: Place a 3M beam over the axles, and then secure the lower axle with a bushing. The orange shaft head on the motor wiggles around, so the 3M beam prevents the gears from separating.

7

x1   x1

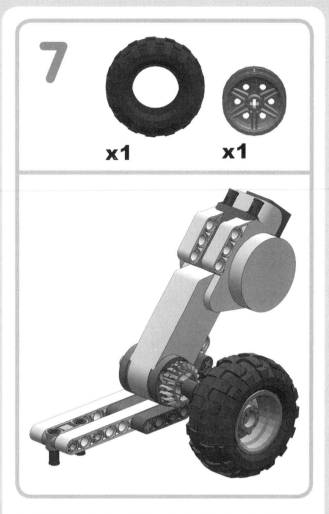

Step 7: Finish by attaching a balloon wheel and a balloon tire. Make sure that the side of the wheel with spokes is facing toward the motor.

right drive subassembly

The *Right Drive subassembly* drives the right side of the robot and, with a few exceptions, mirrors the Left Drive subassembly. There are nine steps.

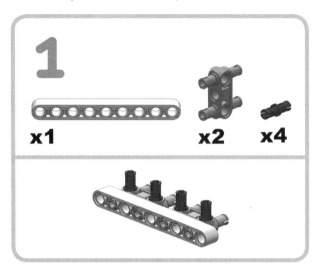

Step 1: Begin by snapping two 3M pegged blocks into a 9M beam and then adding four friction pegs.

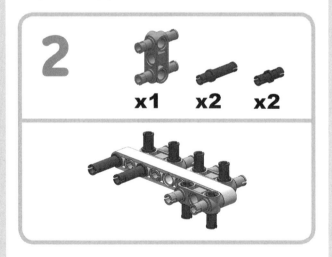

Step 2: Add another 3M pegged block, two 3M friction pegs, and two more friction pegs.

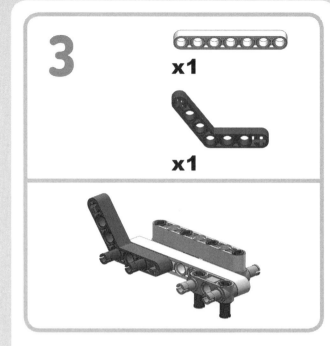

Step 3: Add a 7M beam and a 7M angled beam.

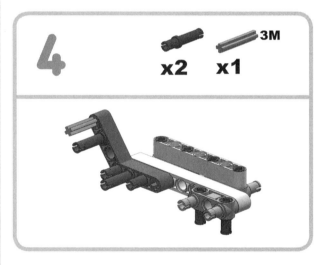

Step 4: Add two 3M friction pegs and a 3M axle to the 7M angled beam.

5

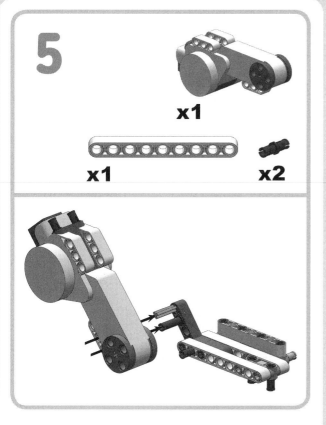

x1

x1   x2

Step 5: Add another 9M beam and a motor to the 7M angled beam. Then connect two friction pegs to the top of the motor.

6

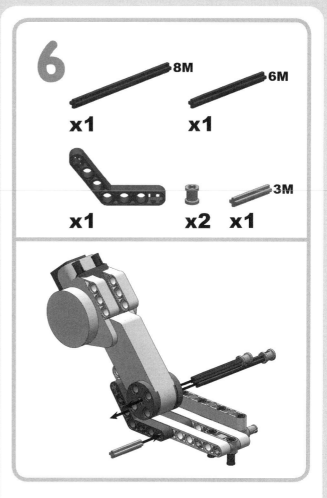

8M   6M

x1   x1

x1   x2   x1   3M

Step 6: First, attach another 7M angled beam. Next, push a 3M axle through the two angled beams. Finally, add bushings to 8M and 6M axles and then push them through as shown, with the 6M axle running through the motor's output shaft and the 8M axle running through the 7M angled beams.

7

x1   x1

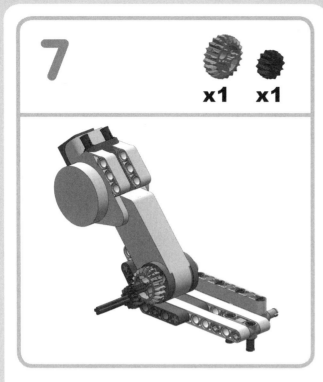

Step 7: Connect a 20t double bevel gear to the 6M axle and a 12t double bevel gear to the 8M axle.

8

x1   x1

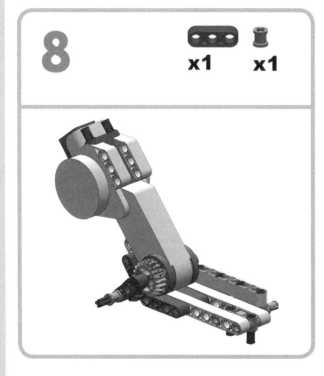

Step 8: Place a 3M beam over the axles, and then secure the lower axle with a bushing.

9

x1   x1

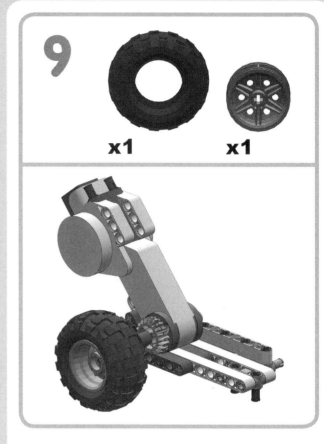

Step 9: Finish by attaching a balloon wheel and a balloon tire. Make sure that the side of the wheel with spokes is facing toward the motor.

steering subassembly

The *Steering subassembly* is a structure that features two wheels and has an axle protruding from the top. This axle connects directly to a motor's output shaft (the motor you'll build in "Steering Motor Subassembly" on page 166) and turns the wheels when rotated by the motor. This subassembly is based on a wheel steering mechanism designed by Bryan Bonahoom for his robot Collision Avoidance System (CAS).[*] There are seven steps.

---

[*] See http://mindstorms.lego.com/MeetMDP/BBonahoom.aspx for more information about CAS.

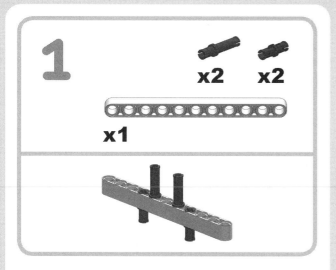

1 x2 x2

x1

Step 1: Begin by snapping two 3M friction pegs and two friction pegs into an 11M beam.

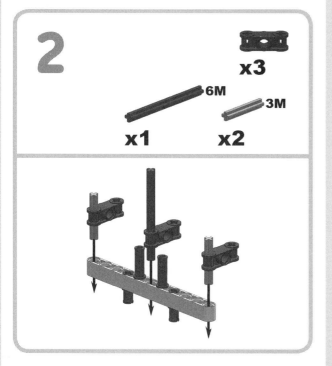

2 x3

6M

3M

x1 x2

Step 2: Add three inverted cross blocks, two 3M axles, and a 6M axle.

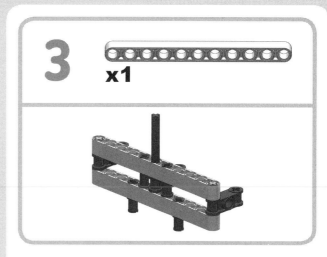

3 x1

Step 3: Push another 11M beam onto the 3M friction pegs and the axles.

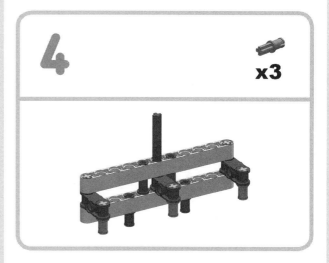

4 x3

Step 4: Flip the subassembly around, and add three axle pegs to the inverted cross blocks.

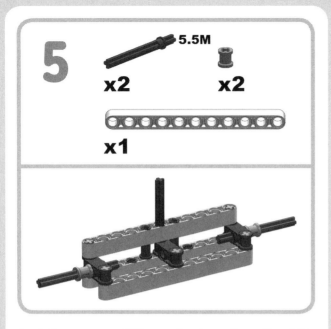

Step 5: First, connect an 11M beam to the axle pegs. Next, add two 5.5M stopped axles to the two outermost inverted cross blocks and secure them each with a bushing. Make sure that the "stop" portion of these axles is on the inside of the subassembly.

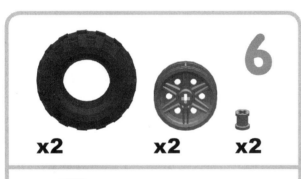

Step 6: Turn the subassembly around again and add the balloon wheels and balloon tires, securing each wheel with a bushing.

Step 7: Finish by adding two 5M pegged perpendicular blocks. You'll use these to help attach this subassembly to the rest of the robot in the final assembly.

steering motor subassembly

The *Steering Motor subassembly* consists of only a few pieces—and a servo motor plays the major role. In the final assembly, the motor in this subassembly will attach directly to the 6M axle in the Steering subassembly. There are only two steps.

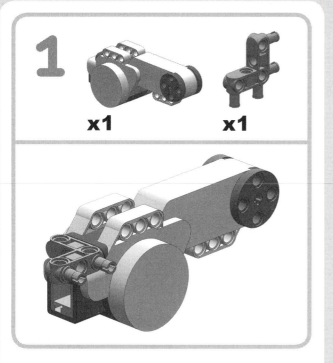

1 x1 x1

Step 1: Push a 5M pegged perpendicular block into the back of a servo motor.

2 x1 x3 x1(35 cm/14 inch)

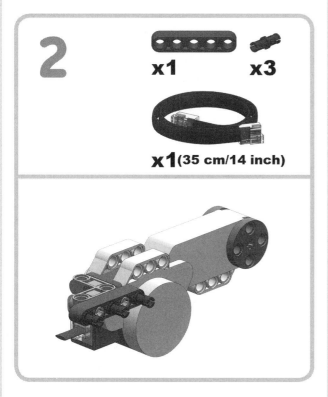

Step 2: Finish by adding a 5M beam, three friction pegs, and a 35 cm/ 14 inch cable. You'll connect the cable to the NXT in the final assembly.

frame subassembly

The *Frame subassembly* mainly provides a place for attaching the NXT in the final assembly. There are four steps.

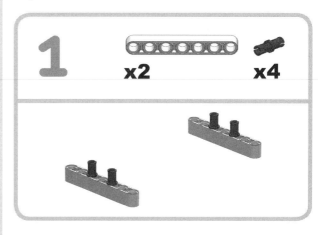

1 x2 x4

Step 1: Begin by laying out two 7M beams and placing two friction pegs in each of them.

2 x4 x4

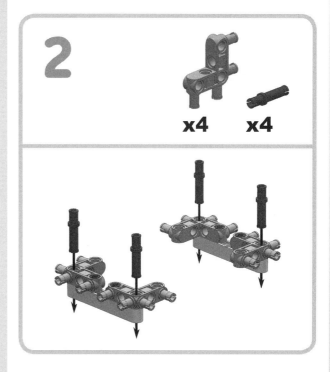

Step 2: Add four 5M pegged perpendicular blocks and push one 3M friction peg through each of them.

3 x4 x2

x2

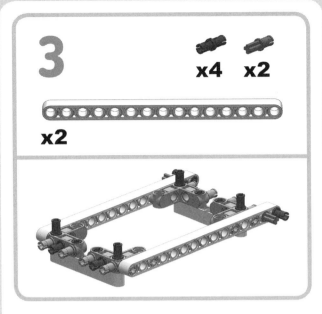

Step 3: Add two 15M beams, four friction pegs, and two friction axle pegs. Note that two of the friction pegs are added to the 5M pegged perpendicular blocks on the front end of the subassembly.

4 x2 x2

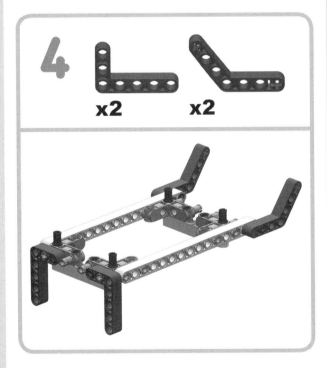

Step 4: Finish by adding two 7M angled beams and two 7M perpendicular angled beams.

ultrasonic sensor subassembly

The *Ultrasonic Sensor subassembly* connects directly to the shaft head on the motor in the Steering Motor subassembly. This means that the motor in the Steering Motor subassembly not only steers the robot but also points the ultrasonic sensor in different directions. In this way, Tag-Bot can look at different paths with the ultrasonic sensor and, based on the sensor's reading, determine which one it wants to take. We'll explore this concept later in the programming (see "Playing Flashlight Tag: A More Advanced Approach" on page 176). There are four steps for this subassembly.

1 x1 x3

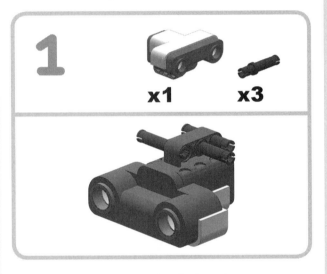

Step 1: Push three 3M friction pegs into the ultrasonic sensor.

2 x2 x2

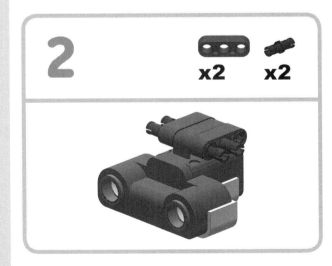

Step 2: Add two 3M beams and two friction pegs.

3

x2   x2   x4

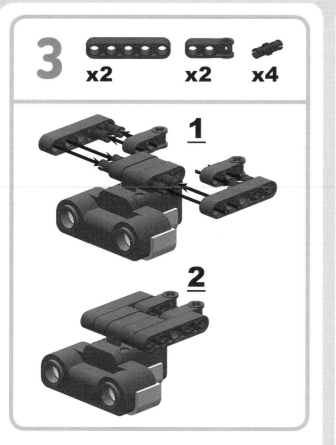

1

2

Step 3: Push four friction pegs into two 5M beams, add an extended cross block to each beam, and then push the 5M beams onto the ultrasonic sensor.

4

x1(20 cm/8 inch)   x2

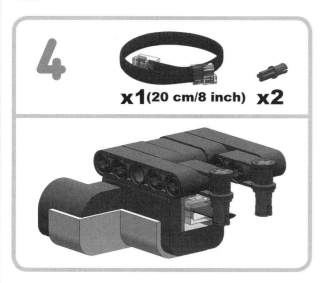

Step 4: Finish by adding two friction axle pegs to the extended cross blocks and plugging in a 20 cm/8 inch cable. You'll connect the cable to the NXT in the final assembly.

light sensor subassembly

The *Light Sensor subassembly* features the light sensor and connects to the NXT in the final assembly. There are three steps.

1

x1   x3

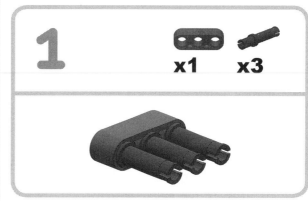

Step 1: Add three 3M friction pegs to a 3M beam.

2

x1

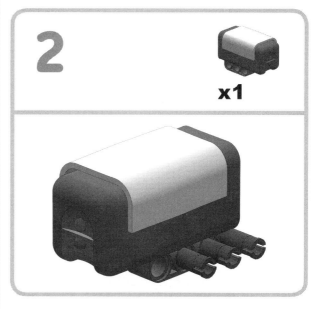

Step 2: Push the light sensor onto two of the 3M friction pegs.

3

x1   x2

Step 3: Finish by adding a 7M perpendicular angled beam and two friction pegs.

sound sensor subassembly

The *Sound Sensor subassembly* also connects to the NXT in the final assembly and features the sound sensor. This subassembly has three steps.

1

x1   x3

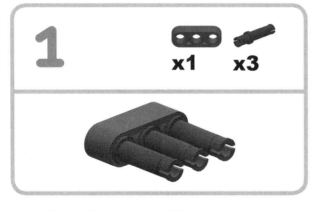

Step 1: Add three 3M friction pegs to a 3M beam.

2

x1

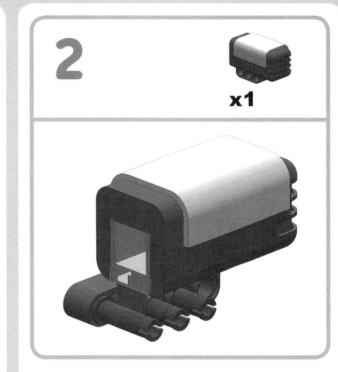

Step 2: Push the sound sensor onto two of the 3M friction pegs.

3

x1   x2

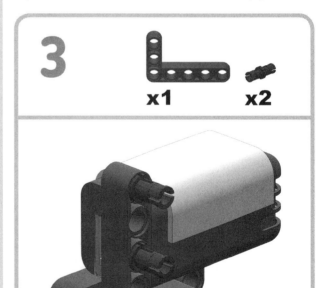

Step 3: Finish by adding a 7M perpendicular angled beam and two friction pegs.

final assembly

With all eight subassemblies built, you're finally prepared to combine them in the final assembly. There are seven steps.

Step 1: Snap together the Left and Right Drive subassemblies.

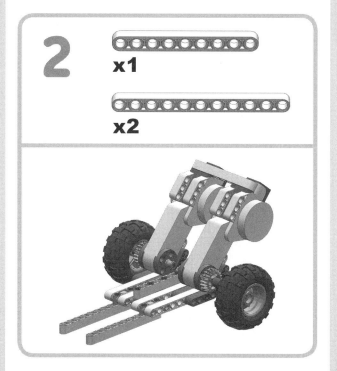

Step 2: Snap a 9M beam onto the tops of the motors, and then attach two 11M beams to the friction pegs on the bottom of the Left and Right Drive subassemblies.

Step 3: Attach the Steering subassembly to the robot.

Step 4: Attach the Steering Motor subassembly to the robot. You'll need to hold a finger under the 6M axle on the Steering subassembly to prevent the motor from pushing it out of place. It's also helpful to pass the motor's electrical cable between the other two motors.

5

x8

Step 5: Attach the Frame subassembly by using eight bushed friction pegs.

6

x1

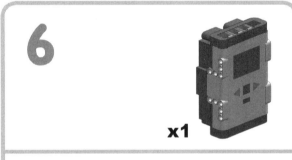

Step 6: Push the NXT onto the friction pegs in the Frame subassembly.

7

x2(35 cm/14 inch)

x2(50 cm/20 inch)

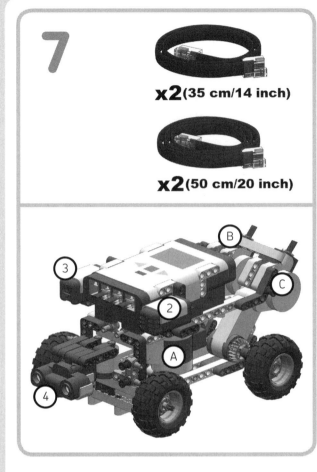

Step 7: Push the Ultrasonic Sensor subassembly onto the steering motor's shaft head and connect its cable to input port 4. Add the Light Sensor and Sound Sensor subassemblies to the NXT, connecting them to the ports shown using 35 cm/14 inch cables. Connect the Steering Motor subassembly's motor to output port A using its cable. Connect the driving motors to the NXT with two 50 cm/20 inch cables as shown.

program-
ming and
testing
tag-bot

Programming Tag-Bot poses certain
challenges. Tag-Bot must not only
determine when a bright light has
shone on its light sensor for more than
a specified amount of time, but it also
must detect and maneuver around
obstacles as it runs around the room.
We'll begin with a basic approach to
accomplishing these tasks. Then we'll
enhance the program by giving Tag-Bot
the ability to look at different paths with
its ultrasonic sensor and choose one
path over another based on how far it
can travel before having to maneuver
around an object.

NOTE Remember that all of the
NXT-G programs used by the robots
in this book are available for download from this book's companion
website (http://www.nxtguide
.davidjperdue.com).

playing flashlight tag:
the basic approach

In our basic program, Tag-Bot1 (Figure 14-3), one of the most important
aspects is its use of a parallel sequence
beam. The upper sequence beam
includes blocks for detecting light, and
the lower sequence beam includes
blocks for detecting objects and
performing random movement. As you
learned in "Parallel Sequence Beams"
on page 76, these sequence beams
execute their programming blocks at
the same time, enabling the robot to
simultaneously perform multiple tasks.
To help us understand this program,
let's look at its individual sections.

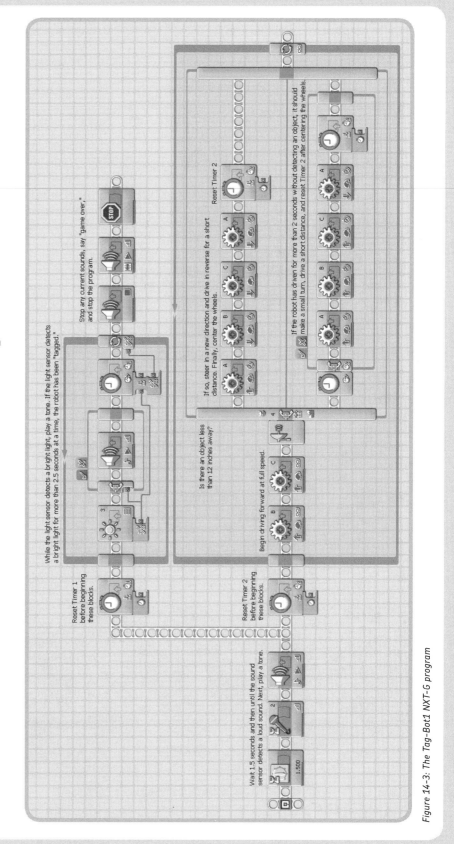

Figure 14-3: The Tag-Bot1 NXT-G program

understanding the tag-bot1 program

The Tag-Bot1 program begins with the three programming blocks shown in Figure 14-4. After waiting 1.5 seconds (enough time for you to step away from the robot), another Wait block instructs the program to wait until the sound sensor detects a sound level greater than 50 percent. To trigger this block, you can clap your hands, at which point a Sound block plays a loud tone, letting you know the main part of the program is about to begin.

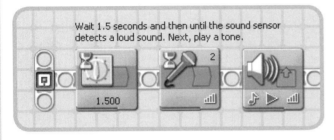

Figure 14-4: The first three blocks in the Tag-Bot1 program

On the upper sequence beam (Figure 14-5), a Timer block first resets Timer 1, which began counting when you ran the program.[*] This timer is used in the following Loop block, which repeats until it receives a *true* logic value via a data plug on its rightmost side. Inside the Loop block is a Light Sensor block that checks the light sensor for a reading less than 25 percent—a condition that's true most of the time (we'll be calibrating the light sensor in a moment). However, when you shine the flashlight on the light sensor, the Light

Sensor block sends a *false* value to the following Switch block and Timer block because the sensor is now reading *more than* 25 percent.

What happens next? The Switch block, configured to read a logic value, plays a tone if its input value is *false* and does nothing if its input value is *true*. Therefore, you'll know that Tag-Bot is detecting enough light from your flashlight when it plays a tone. The following Timer block, however, is the key to the program. It asks, "Does Timer 1 read more than 2.5 seconds?" Since it is receiving logic data on its Reset plug from the Light Sensor block, reaching 2.5 seconds isn't so easy. As long as the Light Sensor block sends the *true* logic value (no bright light is detected), Timer 1 is continually reset. Only when the light sensor detects a bright light can Timer 1 progress. Hence, keeping the flashlight trained on the light sensor for more than 2.5 seconds causes Timer 1 to finally reach and pass 2.5 seconds. At that point, the Timer block sends a *true* value to the Loop block, ending the loop. The robot has been tagged! The following Sound block stops any current tones, and then another Sound block instructs the robot to say, "Game over." A Stop block then abruptly stops the program.

Meanwhile, on the lower sequence beam (Figure 14-6), another Timer block resets Timer 2, and then a Loop block that repeats forever executes. Two Motor blocks immediately turn on motors B and C at full speed. Next, a Switch block asks if the ultrasonic sensor detects an object less than 12 inches away. If it does, the blocks on its top sequence beam execute, causing the robot to steer in a new direction, back up, and then center the wheels again. Finally, Timer 2 is reset. We find out what this timer does next.

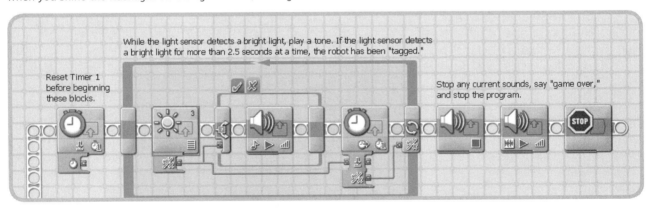

Figure 14-5: The programming blocks on the upper sequence beam

---

[*] I mentioned in Chapter 7 that all Sensor blocks use data wires, but here is an exception: Resetting a timer with the Timer block doesn't require the use of a data wire.

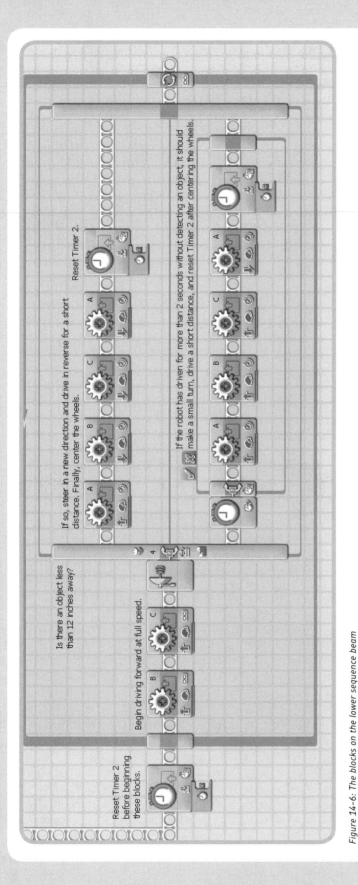

Figure 14-6: The blocks on the lower sequence beam

NOTE Using the Motor block instead of the Move block to drive Tag-Bot is crucial. The Move block would make it much more difficult for Tag-Bot to turn because it attempts to synchronize the motors' movements. With the Motor block, the wheels can slip much more easily during turns.

If the ultrasonic sensor doesn't see an object less than 12 inches away, the blocks on the Switch block's bottom sequence beam execute. On this sequence beam, the first block is *another* Switch block. This Switch block asks if Timer 2 is more than 2 seconds. If it isn't, nothing happens. If it is, the robot steers in a new direction and drives for a small distance before centering the wheels. Finally, it resets Timer 2. In essence, this feature adds some random movement after every 2 seconds of inactivity.

NOTE When rotating motor A (the steering motor), I always use 0.1 rotations. First, this duration achieves the maximum steering angle for Tag-Bot's design. Second, consistently using this duration makes it easy to keep track of the wheels' position because there are only three possibilities: steering left, steering right, or centered.

testing the tag-bot1 program

Now that you've got a grasp on how the Tag-Bot1 program functions, you're ready to test it. First, you need to *calibrate* your light sensor. The primary goal of doing this is to make the light sensor more sensitive to the light of a flashlight. After connecting the NXT to your computer, select **Tools ▸ Calibrate Sensors** from the menu bar to bring up the Calibrate Sensor dialog (Figure 14-7). Make sure that **Light Sensor** is selected in the Sensor Type column, choose **3** for the Port option, and then click the **Calibrate** button on the lower-left corner. The software will immediately download a program called *Calibrate* to the NXT, and the NXT will in turn immediately run the program.

NOTE If you want to return a sensor to its default settings, select the sensor in the Calibrate Sensor dialog and then click the Default button near the bottom of the dialog.

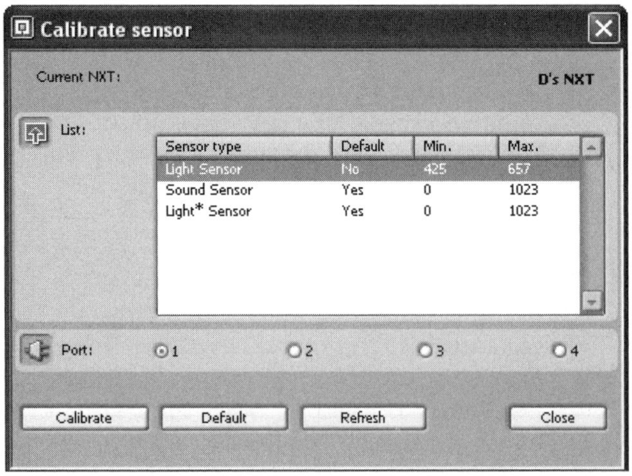

Figure 14-7: You can easily calibrate sensors from the Calibrate Sensor dialog.

With the Calibrate program running, place Tag-Bot somewhere on the floor where there is a moderate level of light. If you look at the LCD on the NXT, it should be displaying the raw value of the light sensor (0–1,023) and asking you to confirm the minimum value (0 percent). Firmly press and release the orange Enter button on the NXT to confirm the minimum value. The LCD should then ask you to confirm the maximum value (100 percent). Take a flashlight and shine it on the robot's light sensor. In addition, you should hold the flashlight above the robot and shine downward, since this is the angle from which you'll most likely be attempting to "tag" the robot. With your other hand, firmly press and release the orange Enter button once again to confirm the maximum value. Depending on how Tag-Bot performs in a moment, you may have to recalibrate the light sensor and experiment with different settings to make it more or less sensitive. In fact, you might need to calibrate the light sensor several times before achieving satisfactory results.

Next, download the Tag-Bot1 program to Tag-Bot. When you're finished, place Tag-Bot on the floor facing away from you, make sure that the wheels are centered (facing straight ahead), and start the program. As usual, make sure the robot has some room to move around. Step back, wait a moment, and then loudly clap your hands. Tag-Bot should begin driving forward—and you should grab a flashlight!

NOTE Whenever you run a program for Tag-Bot, the wheels must be centered because Tag-Bot assumes this is its starting position. If the wheels are in another position, Tag-Bot may not be able to turn the wheels for their full duration, leaving the robot's steering motor in a stall.

As Tag-Bot dashes around the room and continually changes direction, you must continually move around as well in order to shine the flashlight on the robot's light sensor. When Tag-Bot plays a tone, you know that the light sensor is detecting enough light. Although it's possible to shine the flashlight on the robot's back or side and manage to reflect enough light into the sensor, the vast majority of time, you will need to shine the flashlight on the front of the robot. Remember also that you must shine the flashlight on the light sensor for 2.5 seconds *at one time*. The timer is reset whenever the light sensor reads below 25 percent.

When you've finally tagged Tag-Bot, it will eventually say, "Game over" and then coast to a stop. Try playing the game a few more times and even changing the time you must shine the flashlight on the light sensor. To do this, just select the Timer block inside the Loop block on the upper sequence beam and change the Compare parameter from 2.5 to another value. Higher values will make the game more difficult; lower values will make the game easier.

NOTE The current battery level of the NXT and the strength of your flashlight will affect how easily Tag-Bot can be tagged. For best results, use fresh alkaline batteries both in the NXT and in your flashlight.

playing flashlight tag: a more advanced approach

One weakness in the Tag-Bot1 program is that it always initiates the same response when the ultrasonic sensor detects an object (i.e., it instructs Tag-Bot to maneuver around the object in the same way). With this approach, Tag-Bot may end up running into one object as it attempts to maneuver around another one! The enhanced Tag-Bot2 program addresses this problem (Figure 14-8). At first glance, this program looks nearly identical to Tag-Bot1. The changes are mainly within the Switch block that is configured to read an ultrasonic sensor. After detecting an object, two Motor blocks drive Tag-Bot backward a small distance and then execute a My Block called *Path Detection*. Likewise, when there has been two seconds of inactivity, the Switch block beneath it executes the same Path Detection block. Hidden within this My Block, however, is significant complexity. Let's explore the Path Detection block first and then test the program.

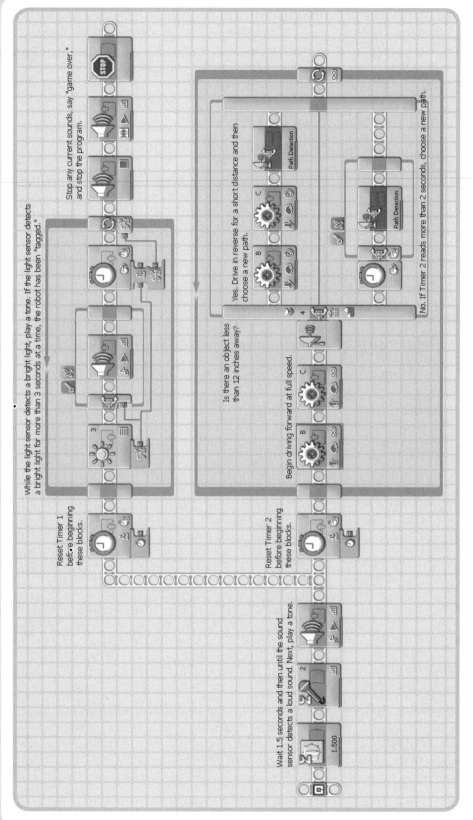

Stop any current sounds, say "game over," and stop the program.

While the light sensor detects a bright light, play a tone. If the light sensor detects a bright light for more than 3 seconds at a time, the robot has been "tagged."

Reset Timer 1 before beginning these blocks.

Is there an object less than 12 inches away?

Yes. Drive in reverse for a short distance and then choose a new path.

No. If Timer 2 reads more than 2 seconds, choose a new path.

Begin driving forward at full speed.

Reset Timer 2 before beginning these blocks.

Wait 1.5 seconds and then until the sound sensor detects a loud sound. Next, play a tone.

Figure 14-8: The Tag-Bot2 NXT-G program

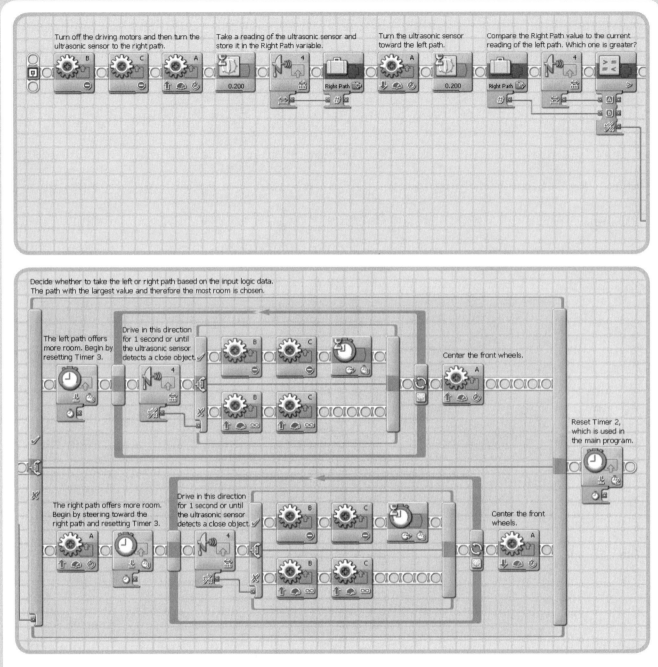

Figure 14-9: The Path Detection block (shown in two parts)

NOTE The Path Detection block must be in the same location on your computer as the Tag-Bot2 program in order for the program to successfully load the block (i.e., the program loads the block from its own directory).

You can, however, add a copy of the Path Detection block to the Custom Palette if you want to use the block in your own programs.

understanding the path detection block

The Path Detection block (Figure 14-9) actually consists of more programming blocks than the main program itself. What does this block accomplish? It enables the robot to look at two possible paths—left and right—by turning its ultrasonic sensor to the farthest left and right positions, and then it instructs the robot to choose which path has more room (i.e., the greatest distance to an object). To help us understand how this program functions, let's break it down into manageable sections.

The first six blocks (Figure 14-10) stop the driving motors, turn the ultrasonic sensor to the rightmost position, wait 0.2 seconds, and then store the ultrasonic sensor's reading in the Right Path variable. The short pause gives the ultrasonic sensor time to stabilize its reading.

The following five blocks (Figure 14-11) turn the ultrasonic sensor to the leftmost position, wait 0.2 seconds for the sensor's readings to stabilize, and then compare the sensor's current reading of the left path to the reading in the Right Path variable. The Compare block asks, "Is the current reading of the left path greater than the reading of the right path stored in the Right Path variable?" If it is, the Compare block sends a *true* logic value. Otherwise, the value is *false*.

Next, a Switch block receives the logic data sent from the Compare block (Figure 14-12). If the input data is *true*, the blocks on the top sequence beam execute, and if the input data is *false*, the blocks on the bottom sequence beam execute. The blocks on the top sequence beam instruct the robot to drive toward the left path for 1 second or until the ultrasonic sensor detects an object that is less than 8 inches away (i.e., the robot is about to run into something).

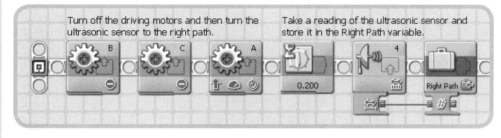

Figure 14-10: The first six blocks in the Path Detection block take a reading of the right path.

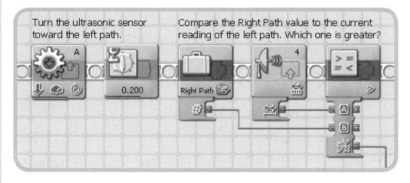

Figure 14-11: These five blocks compare the reading of the left path and the right path.

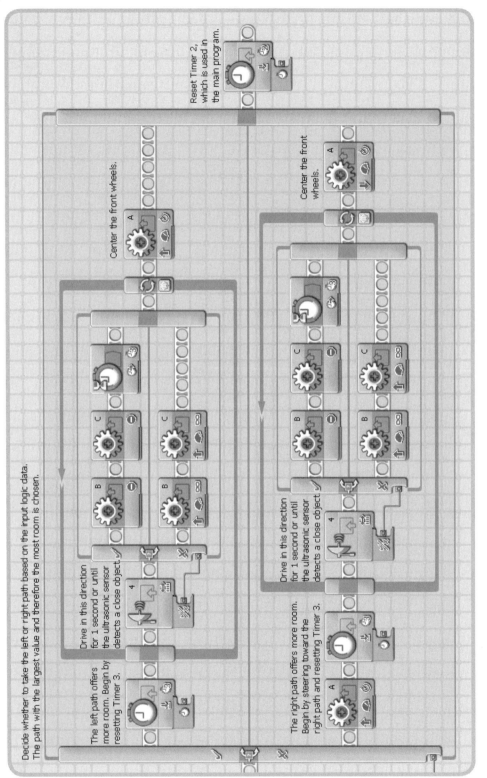

Figure 14-12: This section of the Path Detection block steers the robot for 1 second in the chosen path before driving straight again.

The blocks on the bottom sequence beam instruct the robot to drive toward the right path for 1 second or until the ultrasonic sensor detects an object that is less than 8 inches away. Finally, a Timer block resets Timer 2 for the main program.

testing the tag-bot2 program

Download the Tag-Bot2 program to Tag-Bot, and then position the robot on the floor with its wheels centered as before. Run the program, step back, and then clap your hands to start the main part of the program. As Tag-Bot begins running around, notice how it observes its two potential paths and then chooses one. The downside of this approach is that it takes more time for the robot to make a turn, making Tag-Bot more vulnerable. For this reason, Tag-Bot2 is configured so that you must shine the flashlight on the light sensor for 3 seconds instead of 2.5 seconds. In summary, make sure to play several games, experiment with the blocks' options, and have fun!

NOTE Due to limitations of the ultrasonic sensor, such as its difficulty detecting round objects, Tag-Bot may occasionally choose a path that doesn't offer the most room.

conclusion

Tag-Bot demonstrates not only how to build an NXT vehicle with a steering drive but also how to create a robot that plays a game with a human. In this chapter, you started by constructing eight subassemblies, and then you combined them to create Tag-Bot. You then programmed Tag-Bot to play flashlight tag, beginning with a basic program that enabled the robot to simultaneously detect bright light and obstacles while driving. Finally, you enhanced the program with the Path Detection block, a My Block that enables Tag-Bot to look at two different paths with its ultrasonic sensor and choose one based on the sensor's readings.

guard-bot

In the last few chapters, we've focused on wheeled robots such as Zippy-Bot and Tag-Bot, but another popular design for mobile creations is *legged* robots—robots that walk. While legged robots are not as versatile as wheeled robots, creating a walker is an exciting and rewarding challenge.

In this chapter, we'll create a six-legged robot named Guard-Bot (Figure 15-1) that uses two independent motors to move the legs: One motor powers three legs on the left side, and the other motor powers three legs on the right side. Guard-Bot maintains stability by always having three legs on the ground—two on one side, and one on the other—and Guard-Bot can turn in place by running its legs in opposite directions (similar to how Zippy-Bot can turn in place by driving its wheels in opposite directions). To keep its legs synchronized, Guard-Bot relies on the servo motors' built-in rotation sensors. As its name suggests, however, Guard-Bot doesn't just walk. It can also guard objects or areas, detecting intruders with its ultrasonic sensor and using a ball-launching mechanism on its back end to throw one of the balls from the NXT set. The light sensor on the front of the robot detects the black line on the NXT test pad.

First you'll build this relatively large and complex creation, and then you'll program it to function in two entirely different ways. Specifically, you'll program Guard-Bot to serve as a motion detector that guards a single pathway or object, and then you'll program Guard-Bot to guard a specific area: the NXT test pad.

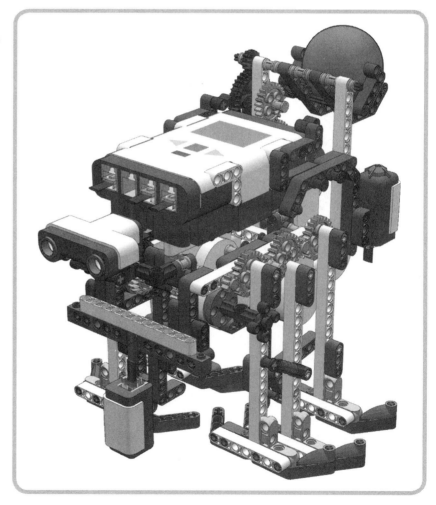

Figure 15-1: Guard-Bot is a six-legged walker that can throw a ball.

building guard-bot

Guard-Bot is composed of eleven subassemblies, but only seven of these are unique (i.e., you'll be building more than one of some subassemblies); Figure 15-2 shows the BOM for Guard-Bot. You'll construct the following seven

subassemblies in the order and quantity shown and then combine them in the final assembly:

* Left Leg subassembly (x3)
* Left Drive subassembly
* Right Leg subassembly (x3)
* Right Drive subassembly
* Ball Launcher subassembly
* Light Sensor subassembly
* Ultrasonic Sensor subassembly

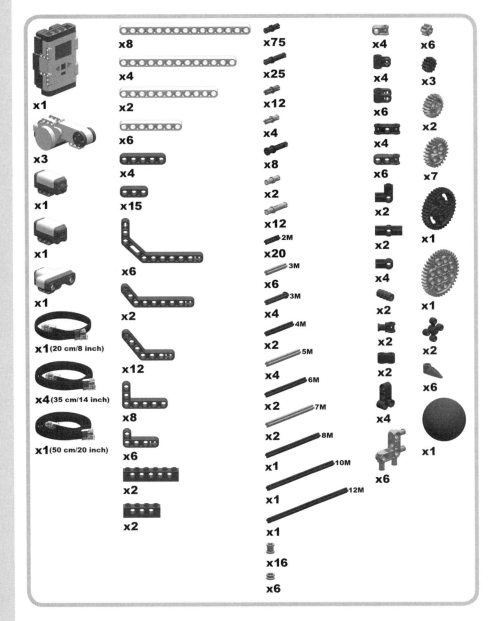

Figure 15-2: A BOM for Guard-Bot

left leg subassembly

The *Left Leg subassembly* is one of Guard-Bot's six legs, and you should follow these building instructions *three times* to build three Left Leg subassemblies. Each one will attach to the Left Drive subassembly that you'll build next (which means you'll be using a subassembly within a subassembly). There are four steps for building one Left Leg subassembly.

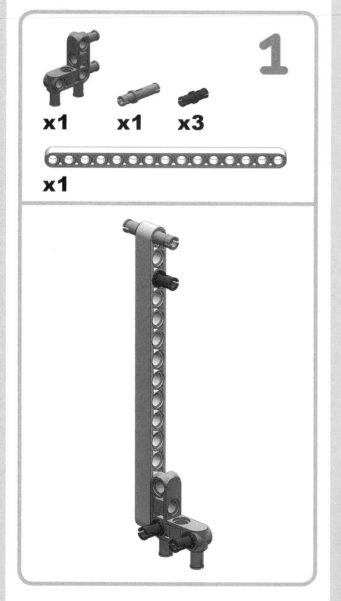

Step 1: Snap a 5M pegged perpendicular block, a 3M peg, and a friction peg into a 15M beam. Then add two friction pegs to the pegged block.

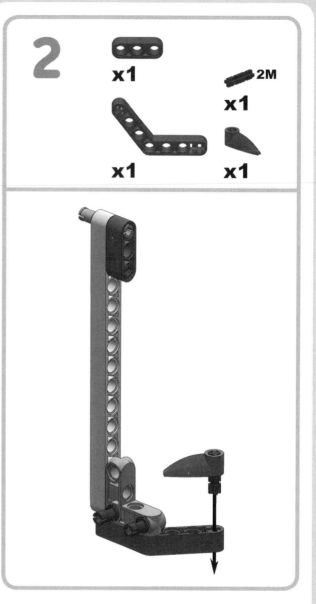

Step 2: Add a 3M beam, a 7M angled beam, and then a TECHNIC tooth along with a 2M notched axle.

3

x1 x2

Step 3: Add a 7M beam and two friction pegs.

4

x1 x1 2M

x1

Step 4: Finish by adding a 7M angled beam, an extended cross block, and a 2M notched axle. Remember to build three Left Leg subassemblies!

4

<u>1</u>

<u>2</u>

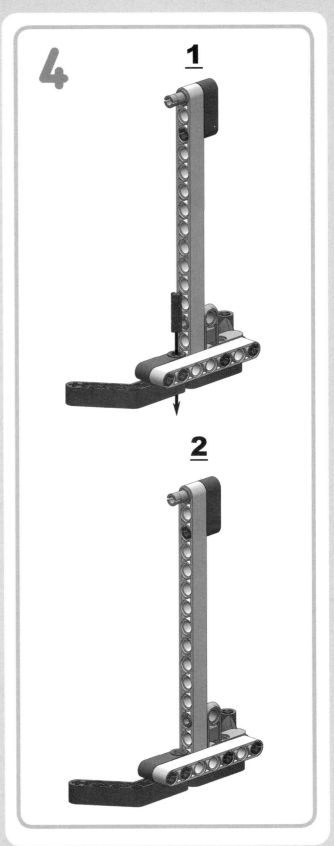

left drive subassembly

The *Left Drive subassembly* moves three legs (the Left Leg subassemblies) on the left side of the robot using a servo motor and a compound gear train. Each leg attaches to a 24t gear that has a gear ratio of 9:5. This is a complex subassembly—the most complex one presented so far in this book—so you should follow the 16 construction steps carefully.

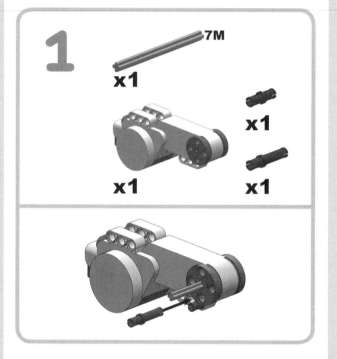

Step 1: Add a 7M axle, a friction peg, and a 3M friction peg to a servo motor.

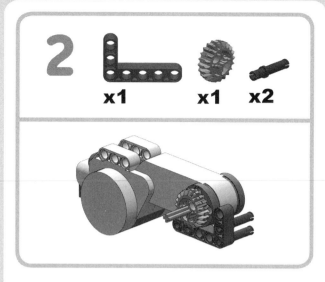

Step 2: Add a 7M perpendicular angled beam, two 3M friction pegs, and a 20t double bevel gear.

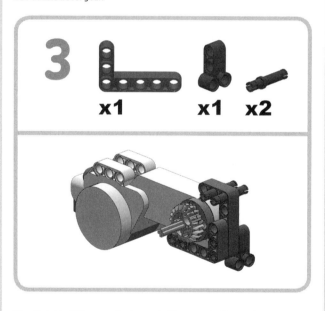

Step 3: Add a 7M perpendicular angled beam, a double peg joiner, and two 3M friction pegs.

4

x1   2M x1

x2   x1

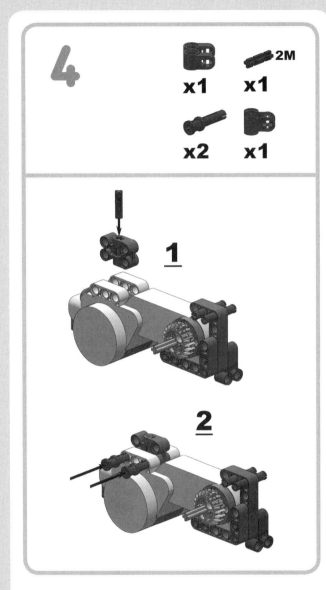

1

2

Step 4: Connect a split cross block and a double cross block with a 2M notched axle, and then secure them to the motor with two bushed friction pegs.

5

7M x1   x1

x1   x1

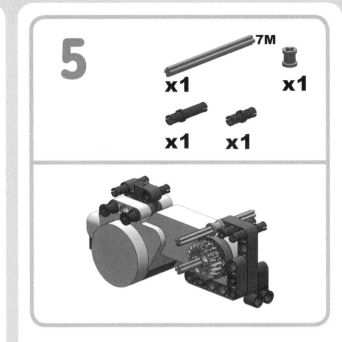

Step 5: Add a 7M axle directly above the motor's shaft head, securing it on one side with a bushing. Also add a friction peg to the split cross block and a 3M friction peg to the double cross block.

6

x2   x1   x1

x1

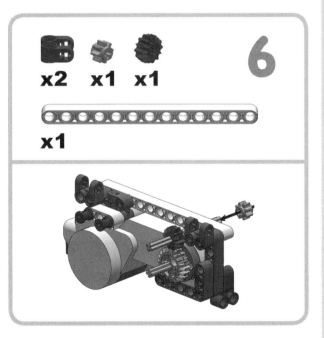

Step 6: Connect a 13M beam to the subassembly, and then add two more split cross blocks. Finally, add an 8t gear and a 12t double bevel gear to the 7M axle as shown.

7

x1 x1 2M

x1 x3

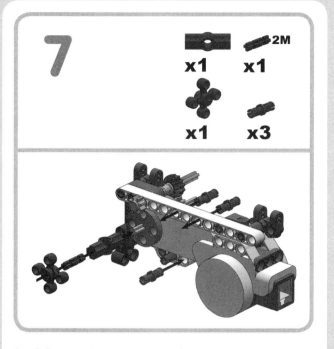

Step 7: Turn the subassembly around to the other side. Add three friction pegs, and then connect a #2 angle connector, a 2M notched axle, and a knob wheel to the motor's output shaft. The knob wheel will make it easy to manually adjust the position of the robot's legs.

8

x1 x1

x2 x2 3M

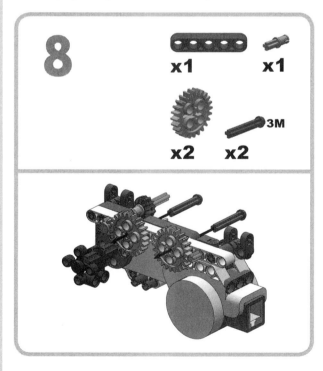

Step 8: Snap a 5M beam onto the two friction pegs on the 13M beam, and then push two 3M studded axles through the 5M beam and into two 24t gears. Also add an axle peg next to the 24t gear on the rightmost side.

9

x1 5M x1 4M

x1 x2 x3

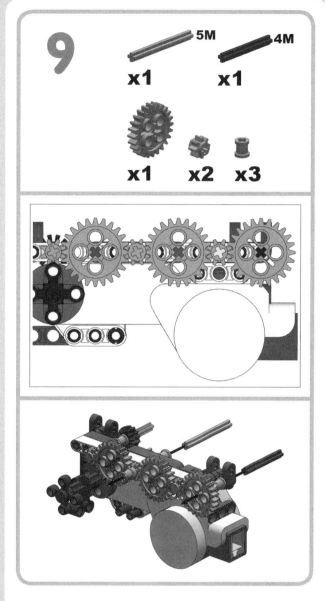

Step 9: Between the two leftmost 24t gears, add a 5M axle along with two bushings and an 8t gear. Next, add an 8t gear to the axle peg. Finally, push a 4M axle through the split cross block, a bushing, and another 24t gear. Make sure that all the 24t gears are positioned exactly as shown.

10

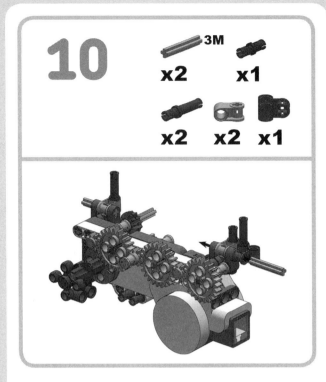

Step 10: Add two 3M axles, two cross blocks, and a double cross block to the split cross blocks. Then snap a friction peg and two 3M friction pegs into the cross blocks and double cross block as shown.

11

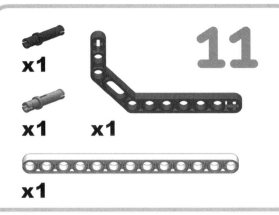

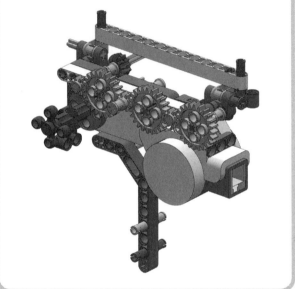

Step 11: On the top, add a 13M beam. On the bottom, add an 11.5M angled beam along with a 3M peg and a 3M friction peg.

12

x2 **x1**

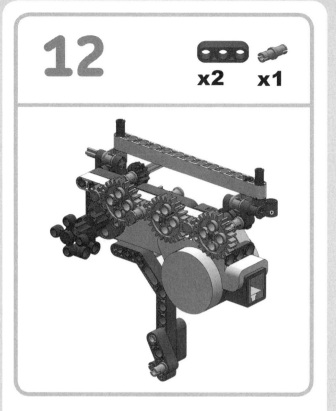

Step 12: Add two 3M beams to the angled beam, and then snap a peg into one of the 3M beams as shown.

13

Step 13: Attach all three Left Leg subassemblies by pushing their 3M pegs into the round-holes on the 24t gears exactly as shown. The leg in the middle also attaches to the peg in the 3M beam.

14

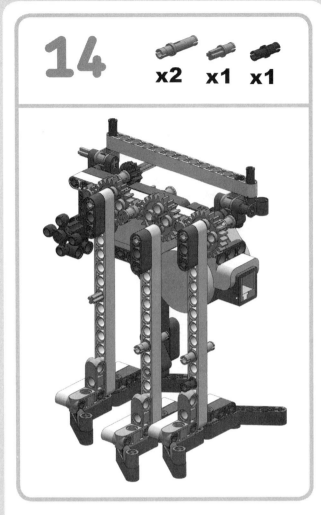

x2   x1   x1

Step 14: Add two 3M pegs, an axle peg, and a friction peg to the legs.

15

x1

x1

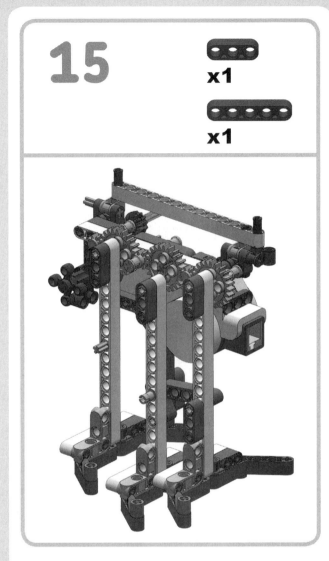

Step 15: Add a 3M beam and a 5M beam to the legs.

16

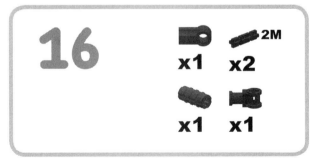

x1   2M x2

x1   x1

Step 16: Using two 2M notched axles, attach a #1 angle connector and a catch to an axle extender. Then attach this to the legs as shown. Congratulations! You're finished with this subassembly.

16

1

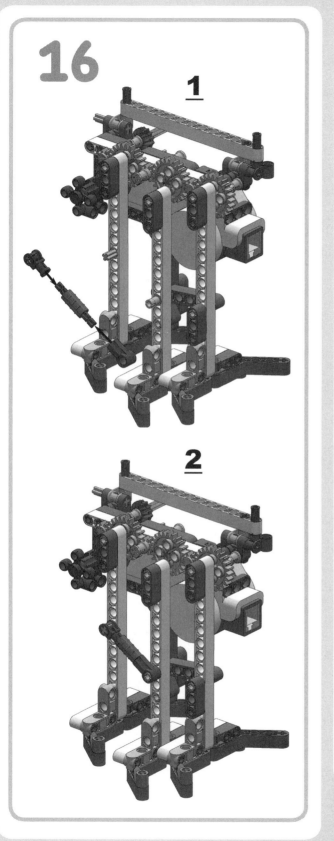

2

right leg subassembly

The *Right Leg Subassembly* is, of course, one of Guard-Bot's six legs and mirrors the Left Leg subassembly. You should follow these building instructions *three times* to build three Right Leg subassemblies, and each one will attach to the Right Drive subassembly that you'll build next.

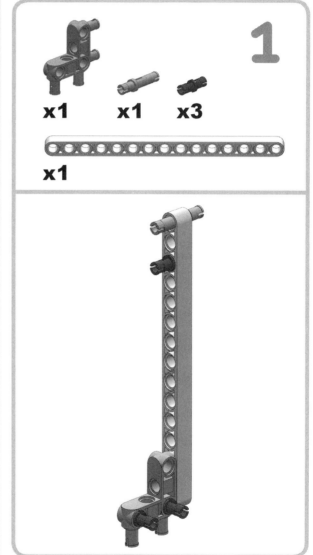

Step 1: Snap a 5M pegged perpendicular block, a 3M peg, and a friction peg into a 15M beam. Then add two friction pegs to the pegged block.

2

x1

x1

2M x1

x1

Step 2: Add a 3M beam, a 7M angled beam, and then a TECHNIC tooth along with a 2M notched axle.

3

x1

x2

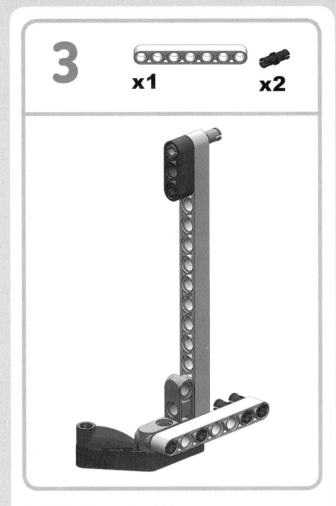

Step 3: Add a 7M beam and two friction pegs.

4

x1

2M x1

x1

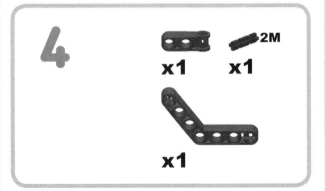

Step 4: Finish by adding a 7M angled beam, an extended cross block, and a 2M notched axle. Remember to build three Right Leg subassemblies!

4

1

2

right drive subassembly

The *Right Drive subassembly* moves three legs (the Right Leg subassemblies) on the right side of the robot using a servo motor and a compound gear train. This subassembly mirrors the Left Drive subassembly with the exception of two pieces (6M axles instead of 7M axles), and there are also 16 construction steps.

1

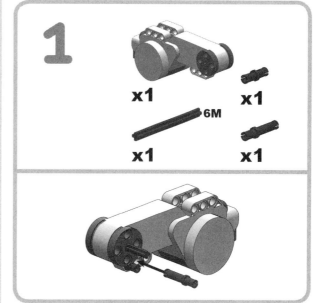

x1   6M
x1   x1
x1   x1

Step 1: Add a 6M axle, a friction peg, and a 3M friction peg to a servo motor.

2

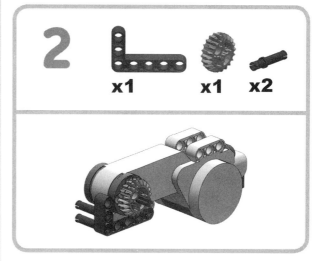

x1   x1   x2

Step 2: Add a 7M perpendicular angled beam, two 3M friction pegs, and a 20t double bevel gear.

GUARD-BOT   **195**

3

x1 x1 x2

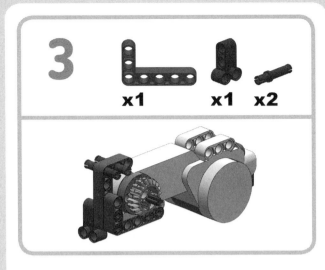

Step 3: Add a 7M perpendicular angled beam, a double peg joiner, and two 3M friction pegs.

4

x1 2M x1 x2 x1

1

2

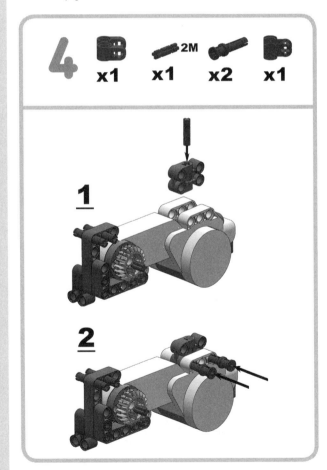

Step 4: Connect a split cross block and a double cross block with a 2M notched axle, and then secure them to the motor with two bushed friction pegs.

5

6M x1 x1

x1 x1

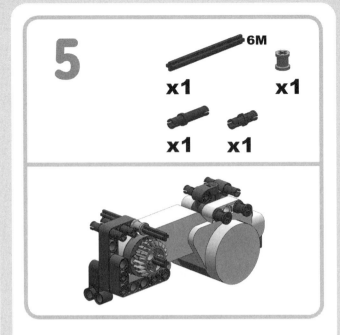

Step 5: Add a 6M axle directly above the motor's shaft head, securing it on one side with a bushing. Also add a friction peg to the split cross block and a 3M friction peg to the double cross block.

6

x2 x1 x1

x1

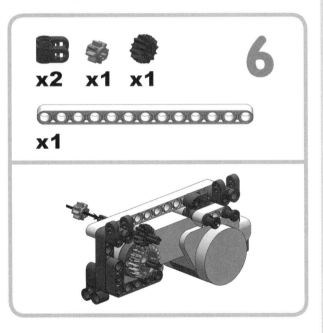

Step 6: Connect a 13M beam to the subassembly, and then add two more split cross blocks. Finally, add an 8t gear and a 12t double bevel gear to the 6M axle as shown.

7

x1 x1 2M

x1 x3

Step 7: Turn the subassembly around to the other side. Add three friction pegs, and then connect a #2 angle connector, a 2M notched axle, and a knob wheel to the motor's output shaft. The knob wheel will make it easy to manually adjust the position of the robot's legs.

8

x1 x1

x2 3M

x2

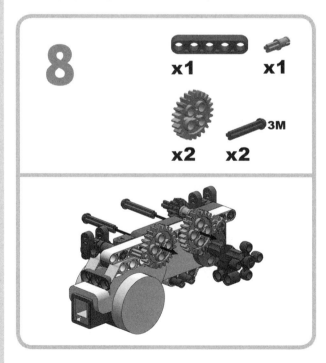

Step 8: Snap a 5M beam onto the two friction pegs on the 13M beam, and then push two 3M studded axles through the 5M beam and into two 24t gears. Also add an axle peg next to the 24t gear on the leftmost side.

9

x1 5M x1 4M

x1 x2 x3

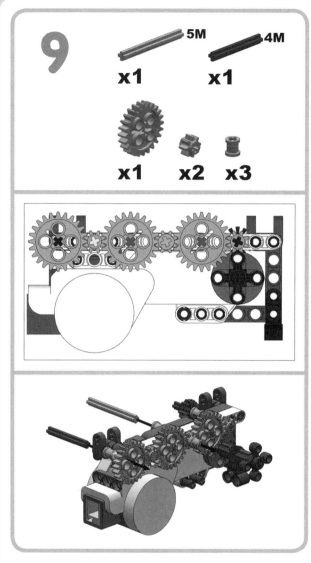

Step 9: Between the two rightmost 24t gears, add a 5M axle along with two bushings and an 8t gear. Next, add an 8t gear to the axle peg. Finally, push a 4M axle through the split cross block, a bushing, and another 24t gear. Make sure that all the 24t gears are positioned exactly as shown.

10

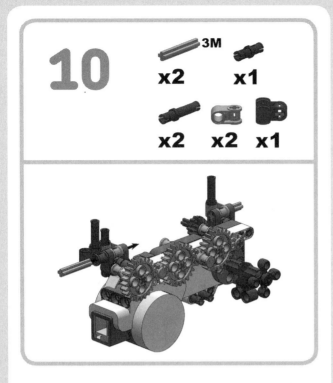

Step 10: Add two 3M axles, two cross blocks, and a double cross block to the split cross blocks. Then snap a friction peg and two 3M friction pegs into the cross blocks and double cross block as shown.

11

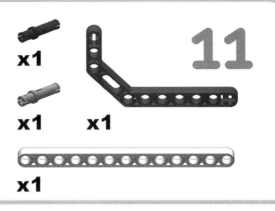

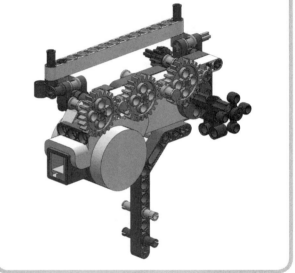

Step 11: On the top, add a 13M beam. On the bottom, add an 11.5M angled beam along with a 3M peg and a 3M friction peg.

12

x2   x1

Step 12: Add two 3M beams to the angled beam, and then snap a peg into one of the 3M beams as shown.

13

Step 13: Attach all three Right Leg subassemblies by pushing their 3M pegs into the round-holes on the 24t gears exactly as shown. The leg in the middle also attaches to the peg in the 3M beam.

14

x2 x1 x1

Step 14: Add two 3M pegs, an axle peg, and a friction peg to the legs.

15

x1

x1

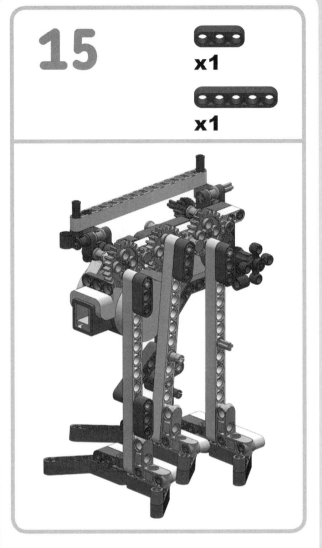

Step 15: Add a 3M beam and a 5M beam to the legs.

16

x1 x1 2M

x1 x1

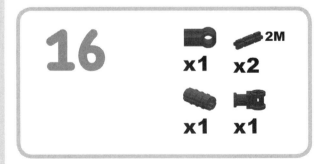

Step 16: Using two 2M notched axles, attach a #1 angle connector and a catch to an axle extender. Finally, attach this to the legs as shown.

16

1

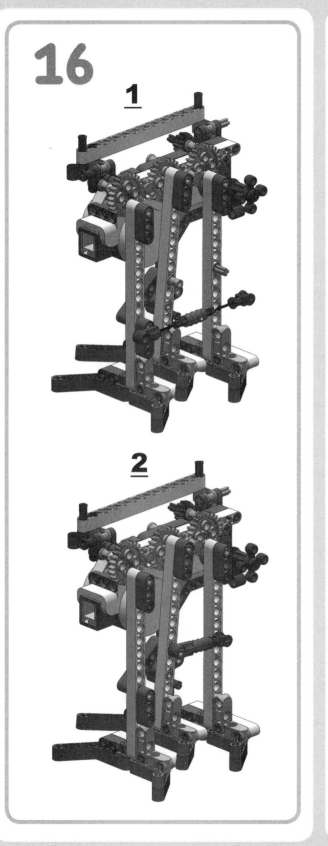

2

ball launcher subassembly

The *Ball Launcher subassembly* uses a servo motor and a compound gear train with a gear ratio of 1:5 to catapult a ball from a hand-like structure. This subassembly can hold and shoot one ball at a time. With fresh batteries in the NXT, this subassembly can easily throw balls a distance of 10 feet! There are 11 construction steps for building this subassembly.

Step 1: Add two friction pegs, two 3M friction pegs, and a 10M axle to a servo motor.

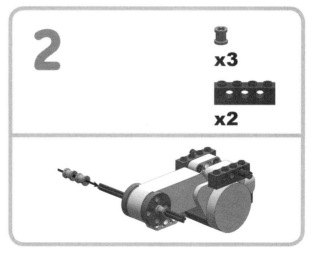

Step 2: Add two 1 × 4 TECHNIC bricks and three bushings. Make sure that you push the bushings all the way to the motor's shaft head.

3

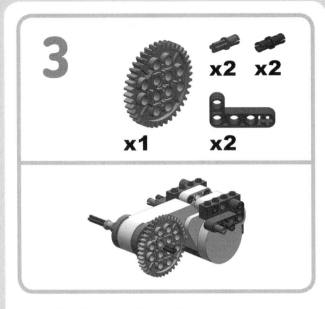

Step 3: Add a 40t gear, two 5M perpendicular angled beams, two friction pegs, and two friction axle pegs.

4

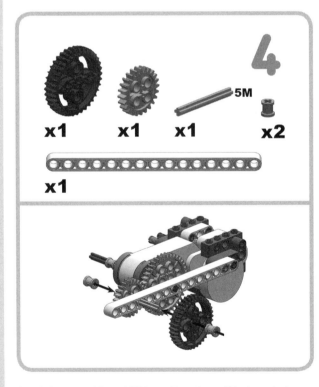

Step 4: Begin by adding a 15M beam. Then place a 5M axle on the beam and push a 24t gear, a 36t double bevel gear, and two bushings onto the axle.

5

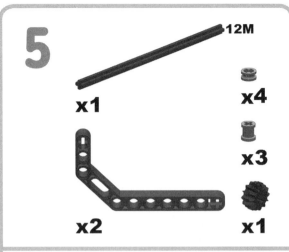

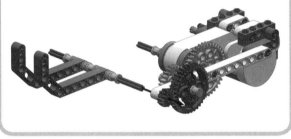

Step 5: Begin by adding three bushings, four half-bushings, and two 11.5M angled beams to a 12M axle. Then push everything through the 15M beam and a 12t double bevel gear.

6

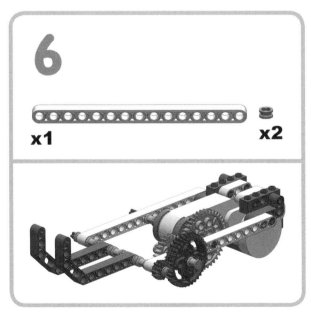

Step 6: Add another 15M beam and two half-bushings to secure the 12M axle.

7

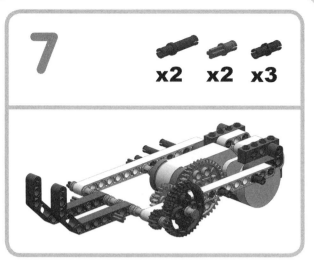

x2   x2   x3

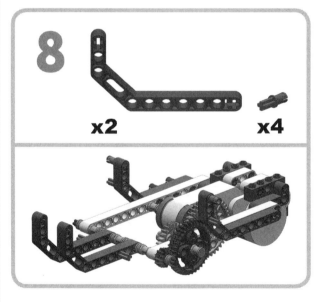

Step 7: Add three friction pegs and two 3M friction pegs to the straight beams as shown, and then add two friction axle pegs to the 11.5M angled beams.

8

x2   x4

Step 8: Add two 11.5M angled beams, and then add four friction axle pegs to those beams.

9

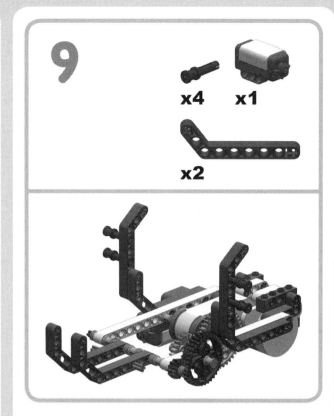

x4   x1

x2

Step 9: Add a touch sensor, two 9M angled beams, and four bushed friction pegs. You'll use these bushed friction pegs in the final assembly to connect the Ball Launcher subassembly to the NXT.

10

8M

x1   x2

Step 10: Add an 8M axle and two inverted cross blocks to the leftmost angled beams.

11 x2

x1 (35 cm/14 inch)

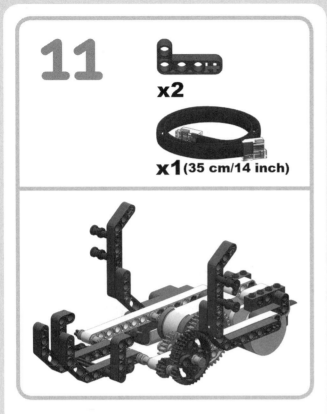

Step 11: Finish by adding two 5M perpendicular angled beams and connecting a 35 cm/14 inch cable to the motor. You'll connect this cable to the NXT in the final assembly.

light sensor subassembly

The *Light Sensor subassembly* attaches to the front of the robot and positions a light sensor so that it faces downward. This light sensor detects the black line on the NXT test pad, but the subassembly itself also connects the two drive subassemblies in the final assembly. There are five construction steps for this subassembly.

1 x1 x2

Step 1: Add two 3M friction pegs to the light sensor.

2 x2 x4

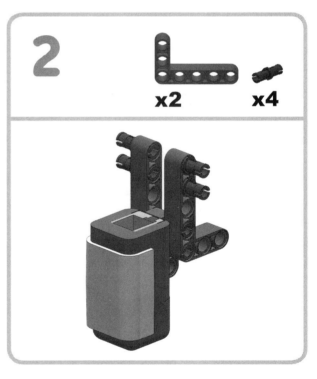

Step 2: Add two 7M perpendicular angled beams and four friction pegs.

3

x2   x4

Step 3: Add two double peg joiners and four more friction pegs.

4

x2   x4

x1

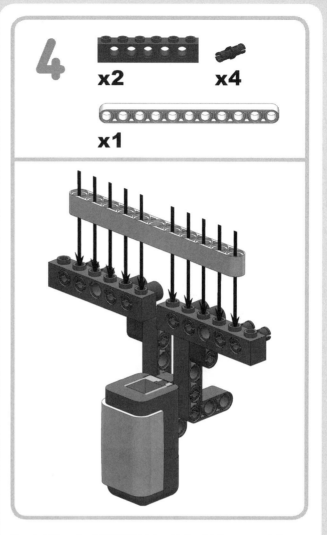

Step 4: Add two 1 × 6 TECHNIC bricks with four friction pegs, and then push an 11M beam onto the bricks' studs.

5

x1 (20 cm/8 inch)

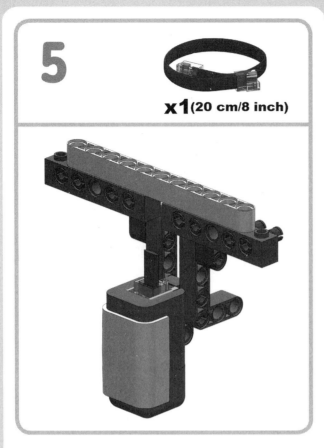

Step 5: Finish by attaching a 20 cm/8 inch cable to the light sensor. You'll connect this cable to the NXT in the final assembly.

ultrasonic sensor subassembly

The *Ultrasonic Sensor subassembly* attaches to the front of the robot in the final assembly and, of course, provides Guard-Bot with its sense of sight. There are only two construction steps.

1

5M x1   3M x1

x1   x2

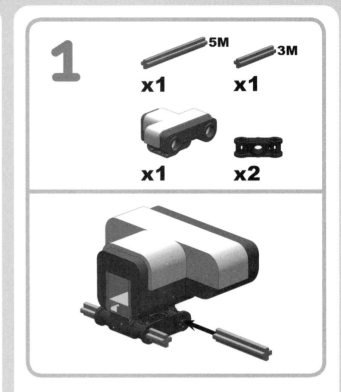

Step 1: Add two inverted cross blocks to the ultrasonic sensor by running a 3M and a 5M axle through them.

2

x1 (35 cm/14 inch) x2

Step 2: Finish by adding two #6 angle connectors and attaching a 35 cm/14 inch cable to the ultrasonic sensor. You'll connect this cable to the NXT in the final assembly.

final assembly

You're almost finished! For the final assembly, there are 12 construction steps.

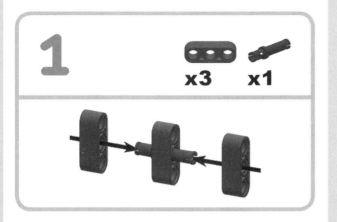

Step 1: Connect three 3M beams to a 3M friction peg.

Step 2: Push the Left and Right Drive subassemblies' axles into the 3M beams; these beams keep the double bevel gears from separating. You'll secure the subassemblies to each other in the following steps.

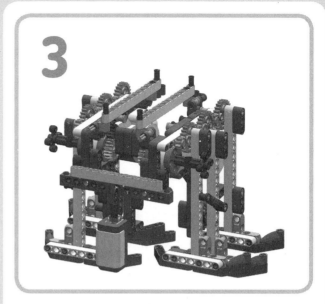

Step 3: Attach the Light Sensor subassembly to the front of the robot, pushing its friction pegs into the double peg joiners.

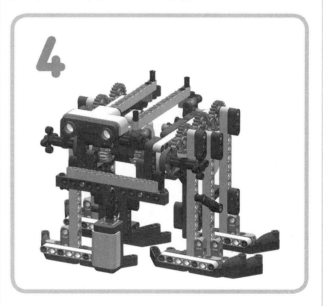

Step 4: Attach the Ultrasonic Sensor subassembly to the front of the robot, pushing its #6 angle connectors into the 3M axles.

5

x6

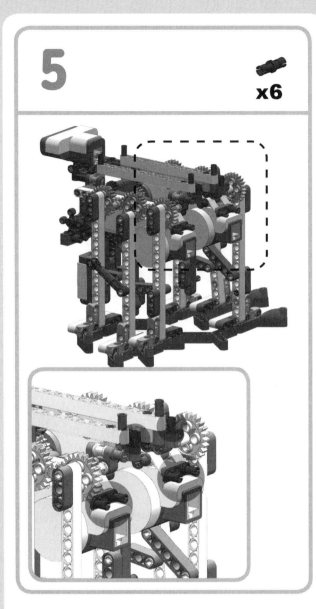

Step 5: Turn the robot around to the other side, and then add six friction pegs: four to the backs of the motors, and two to the double cross blocks on the top.

6

3M

x1   x2

x1

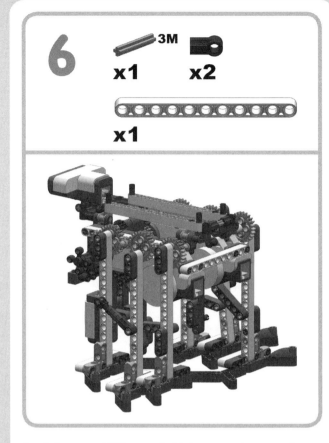

Step 6: Connect an 11M beam to the friction pegs on the motors. Also attach two #2 angle connectors and a 3M axle to the friction pegs on the double cross blocks on the top.

7

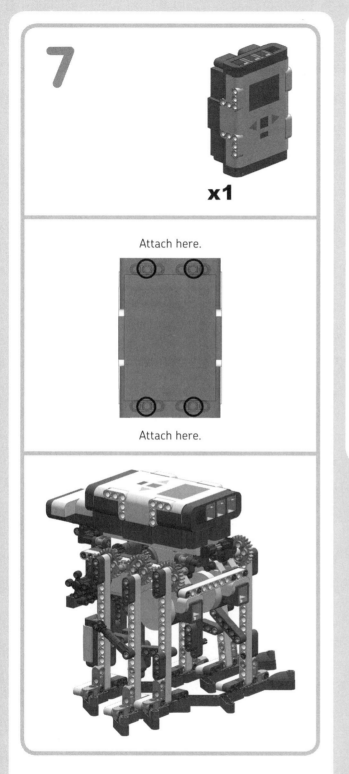

x1

Attach here.

Attach here.

Step 7: Add the NXT by pushing it downward onto the friction pegs. This can be a little tricky, so you should push slowly and focus on one end at a time.

8

x6   **x2**

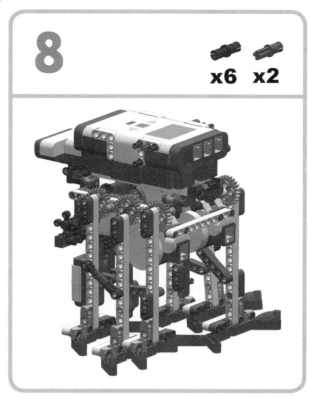

Step 8: Add a friction axle peg and three friction pegs to each side of the NXT.

9

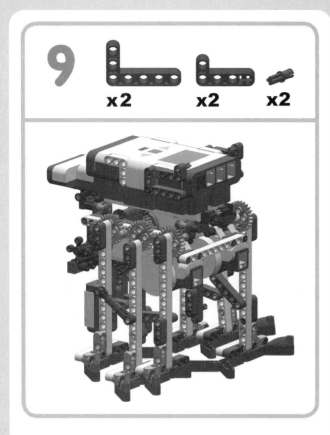

x2   x2   x2

Step 9: Add two 7M perpendicular angled beams and two 5M perpendicular angled beams to the NXT, and then add two friction axle pegs to the 5M perpendicular angled beams.

10

Step 10: Add the Ball Launcher subassembly by pushing its bushed friction pegs into the beams on the NXT. Make sure that the "hand" is in its resting position; it can move backward until one of its inverted cross blocks hits the 24t gear.

11

x1 **x2**

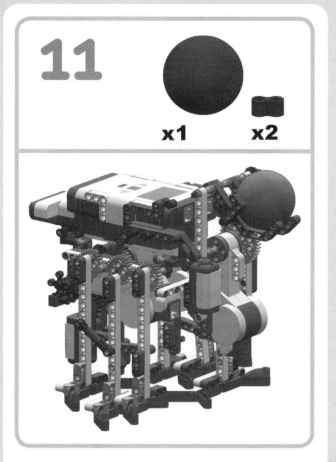

Step 11: Add two flexible axle joiners to the 5M perpendicular angled beams on the NXT, and then load the Ball Launcher mechanism with a ball from the NXT set. The flexible axle joiners help to absorb some of the impact of the hand when it throws the ball.

12

x2 (35 cm/14 inch)

x1 (50 cm/20 inch)

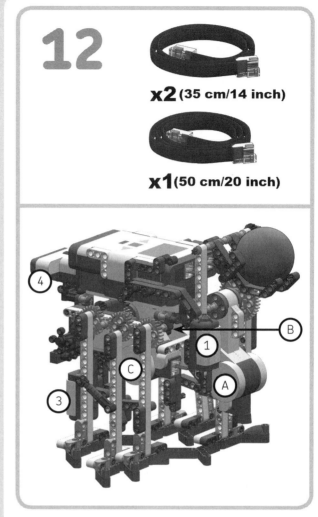

Step 12: Attach a 50 cm/20 inch cable to the touch sensor, and then connect it to port 1. Connect 35 cm/14 inch cables to the Left and Right Drive subassembly motors, and then connect the Left Drive subassembly's motor to port C and the Right Drive subassembly's motor to port B. Connect the other electronic elements' cables to the NXT as shown.

NOTE For the cables connecting to the output ports, make sure that you run them *underneath* the hand on the Ball Launcher subassembly to prevent the hand from getting caught on them.

programming and testing guard-bot

For Guard-Bot, the primary programming challenge is keeping its legs synchronized. To begin, however, we'll program Guard-Bot as a stationary motion detector that guards a single pathway or object. Then we'll program Guard-Bot to walk around the NXT test pad and search for intruders that come within a certain distance. In both scenarios, Guard-Bot puts its ball-launching mechanism to use!

guard-bot: the motion detector

Figure 15-3 shows our first program, Guard-Bot1, in which Guard-Bot simply sits still and detects any motion with its ultrasonic sensor. How does this program give Guard-Bot the ability to detect motion? Using the Range block is the key, but let's break down this program into sections and observe each part to clearly understand it.

The first six blocks execute only once and set the stage for the rest of the program (Figure 15-4). To begin, a Wait block instructs the program to wait one second for the ultrasonic sensor reading to stabilize. Next, an Ultrasonic Sensor block sends the current reading of the ultrasonic sensor to two Math blocks. The first Math block adds 2 to the input value and stores the resulting number in the Upper Bound variable (sixth block). The second Math block subtracts 2 from the input value and stores the resulting number in the Lower Bound variable (fifth block). In a moment, you'll learn the purpose of these variables.

Following the Variable blocks is a Loop block that repeats forever, and the first four blocks inside continually compare the current reading of the ultrasonic sensor with the values stored earlier (Figure 15-5). A Range block uses the Lower Bound variable for its lower bound, the Upper Bound variable for its upper bound, and the current ultrasonic sensor reading as its test value. If the ultrasonic sensor reading is *outside* of the lower or upper bounds, the robot knows it has detected motion. In other words, if the reading is less than the lower bound, something has moved into its line of sight; and if the reading is greater than the upper bound, someone has removed the object Guard-Bot was watching. Once either of these scenarios has occurred, the Range block sends a *true* logic value to the next block.

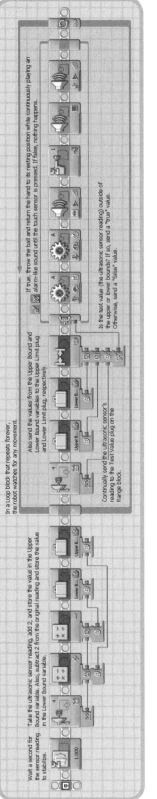

Figure 15-3: The Guard-Bot1 program

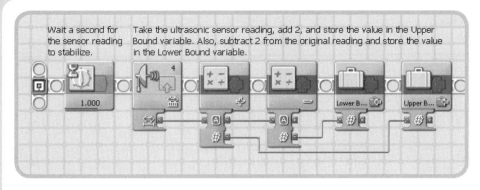

Wait a second for the sensor reading to stabilize.

Take the ultrasonic sensor reading, add 2, and store the value in the Upper Bound variable. Also, subtract 2 from the original reading and store the value in the Lower Bound variable.

Figure 15-4: The first six blocks in the Guard-Bot1 program

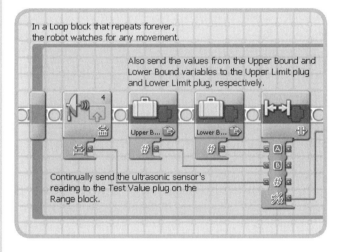

In a Loop block that repeats forever, the robot watches for any movement.

Also send the values from the Upper Bound and Lower Bound variables to the Upper Limit plug and Lower Limit plug, respectively.

Continually send the ultrasonic sensor's reading to the Test Value plug on the Range block.

Figure 15-5: The first four blocks inside the Loop block

NOTE Why does the program subtract 2 from the Lower Bound variable and add 2 to the Upper Bound variable? It's not uncommon for the ultrasonic sensor's readings to slightly vary at times, and expanding the upper and lower bounds in this manner helps to prevent the robot from reacting to these slight changes.

The following block is a Switch block that asks if the input value from the Range block is *true* or *false* (Figure 15-6). If it's *false* (nothing detected), then nothing happens. But if it's *true*, the six blocks shown execute. Two Motor blocks throw the ball and then return the hand in the Ball Launcher subassembly to its resting position. Finally, an alarm-like sound (Error 02) loudly and continuously plays until you've reloaded the ball and pressed the touch sensor. At that point, the alarm stops and a tone plays to confirm that the touch sensor has been pressed. Of course, the Loop block then repeats.

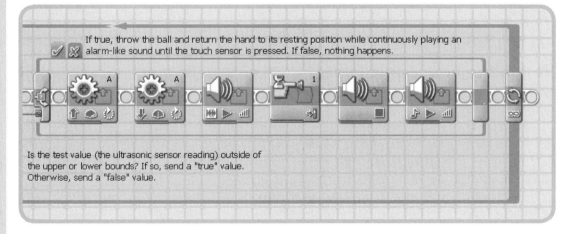

If true, throw the ball and return the hand to its resting position while continuously playing an alarm-like sound until the touch sensor is pressed. If false, nothing happens.

Is the test value (the ultrasonic sensor reading) outside of the upper or lower bounds? If so, send a "true" value. Otherwise, send a "false" value.

Figure 15-6: The Switch block executes six blocks if its input value is true.

Let's test this program. After downloading the Guard-Bot1 program to Guard-Bot, place the robot so that it faces a large, flat object (like a wall) and then start the program. Guard-Bot should be completely quiet. Now try to creep past the ultrasonic sensor's line of sight. As soon as you step into view, Guard-Bot should launch its ball (look out!) and sound its alarm. Pick up the ball, place it back in Guard-Bot's ball-launching mechanism, and press the touch sensor. Guard-Bot should instantly become quiet and resume waiting and watching.

NOTE If Guard-Bot is facing an object that is not flat or standing too far away from an object, the initial ultrasonic sensor reading may not be accurate and may therefore cause Guard-Bot to erratically throw its ball and sound its alarm.

After you've played around for a little while with this setup, stop the program and place Guard-Bot in front of a relatively large but flat object that you can easily move (such as a cardboard box). Run the program and then pick up the object, moving it somewhere else. Guard-Bot should immediately launch its ball and sound its alarm, recognizing that the object has been moved. (Interestingly, you could even pick up Guard-Bot and get the same results.) Place the object exactly where it was before, and press Guard-Bot's touch sensor after reloading its ball. Guard-Bot should now be quiet as it continues watching the object.

NOTE Make sure that none of the electrical cables get in front of the ultrasonic sensor and interfere with its readings. If you need to, try tucking the cables behind the sensor.

Although this program demonstrates some useful programming techniques and Guard-Bot's ball-throwing capabilities, it doesn't actually get the robot walking. So let's move on to a program that does move the robot.

guard-bot: the intruder detector

In our second program, Guard-Bot2 (Figure 15-7), there are three main sections that execute simultaneously by using parallel sequence beams. On the upper sequence beam, the code blocks instruct Guard-Bot how to look for intruders and throw its ball. On the middle sequence beam, the code blocks instruct Guard-Bot how to walk around the NXT test pad, turn around when it detects the black line, and wait for its ball to be reloaded. On the lower sequence beam, the code blocks continuously update the LCD with the number of balls that Guard-Bot has thrown. We'll individually consider each of these three divisions, starting at the top and working our way downward.

NOTE The Guard-Bot2 program includes four different My Blocks. In order for the program to successfully load the blocks, they must be in the same location on your computer as the Guard-Bot2 program (i.e., the program loads the blocks from its own directory). You can, however, add copies of these My Blocks to the Custom Palette if you want to use them in your own programs.

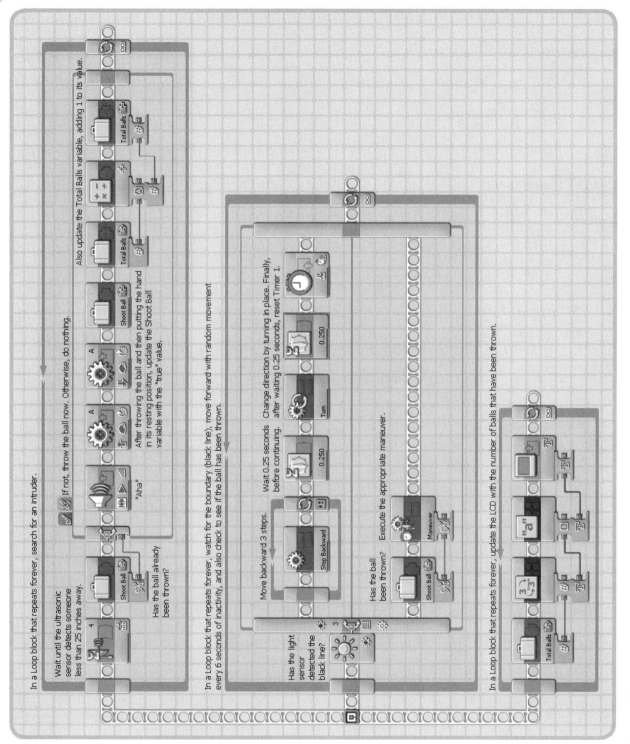

Figure 15-7: The Guard-Bot2 program

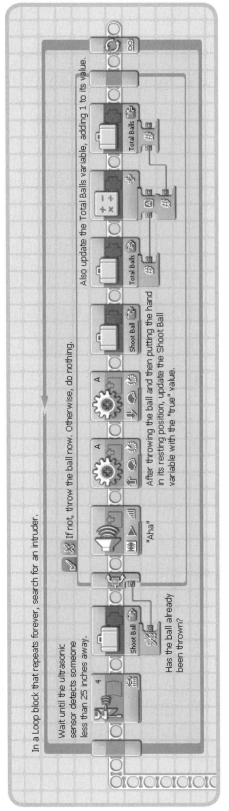

Figure 15-8: The code blocks on the upper sequence beam

The first block on the upper sequence beam is a Loop block set to repeat forever (Figure 15-8). Inside the Loop block, a Wait block asks if the ultrasonic sensor sees anyone less than 25 inches away. When that condition is satisfied, the Shoot Ball variable sends its logic data to a Switch block. The Shoot Ball variable helps the robot keep track of whether it has already thrown a ball. If the value is *true* (the ball has already been thrown), nothing happens; Guard-Bot is waiting for its ball to be reloaded. If the value is *false*, the following seven blocks execute. First, a Sound block instructs the robot to say, "Aha!" Two Motor blocks subsequently power the Ball Launcher subassembly, launching the ball and then bringing the hand back to its resting position. Next, the Shoot ball variable is given the *true* value. Finally, the Total Balls variable is incremented by 1 using a Math block. The code blocks on the lower sequence beam, which we'll examine soon, use the Total Balls variable.

The first block on the middle sequence beam is also a Loop block set to repeat forever (Figure 15-9). Inside the Loop block, a Switch block immediately decides whether the light sensor has detected the line; specifically, it asks if the light sensor reads a value less than 45 percent (you may need to adjust this value depending on the lighting in your room). If it has detected the line, the Switch block's top sequence beam is taken. The robot proceeds to walk three steps backward by using the Step Backward block inside a Loop block that repeats three times. Next, after waiting 0.25 seconds, the robot changes direction with the Turn block, which runs Guard-Bot's legs in opposite directions for 4 steps. Finally, after waiting another 0.25 seconds, a Timer block resets Timer 1; we'll see what this timer is used for in a moment. The *Step Backward block* (Figure 15-10) and the *Turn block* (Figure 15-11) are My Blocks that each use two Motor blocks on parallel sequence beams in order to keep Guard-Bot's legs synchronized. With Guard-Bot's specific walking configuration, using two Motor blocks in this manner works better than using a single Move block, which doesn't provide the same level of control over the motors.

NOTE To help keep its legs synchronized, Guard-Bot always moves in increments of 1 step, which is 648 degrees or 1.8 rotations.

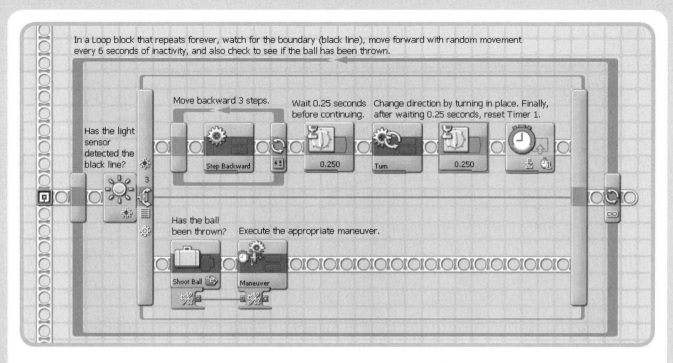

Figure 15-9: The code blocks on the middle sequence beam

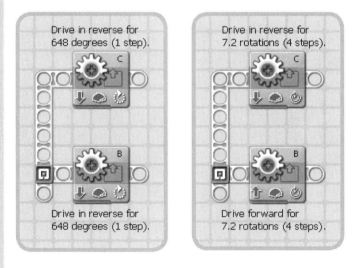

Figure 15-10: The Step Backward My Block instructs Guard-Bot to take one step backward with both sets of legs.

Figure 15-11: The Turn My Block instructs Guard-Bot to change direction by running its legs in opposite directions for four steps.

If the light sensor does *not* detect the line, the bottom sequence beam on the Switch block is taken and the Shoot Ball variable sends its logic data to another My Block: the Maneuver block (Figure 15-12). Inside the Maneuver block, the logic data is received by a Switch block, and if the logic data is *true*, the program waits until the touch sensor has been pressed and released (i.e., waits for you to reload the ball). Finally, a variable block gives the Shoot Ball variable a *false* value, and a Timer block resets Timer 1.

If the logic data is *false* (the ball hasn't been thrown), then *another* Switch block asks if Timer 1 is greater than 6 seconds. If it is, Guard-Bot turns in place using the Turn block, thereby achieving random movement every 6 seconds of inactivity. If the timer hasn't reached 6 seconds, the robot executes the Step Forward block. The *Step Forward block* (Figure 15-13), yet another My Block, is identical to the Step Backward block except that it advances Guard-Bot forward one step. However, because the code for checking the light sensor, the Shoot Ball variable, and the timer executes so quickly, the robot will appear to be taking continuous steps.

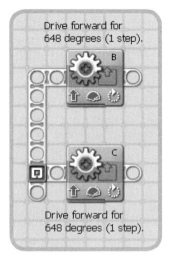

Drive forward for 648 degrees (1 step).

Drive forward for 648 degrees (1 step).

Figure 15-13: The Step Forward My Block instructs Guard-Bot to take one step forward with both sets of legs.

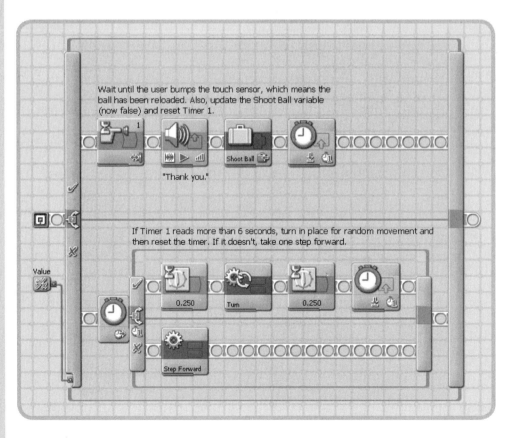

Figure 15-12: The Maneuver My Block receives logic data from the Shoot Ball variable.

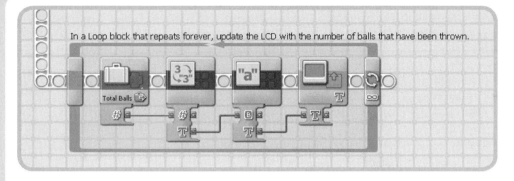

In a Loop block that repeats forever, update the LCD with the number of balls that have been thrown.

Figure 15-14: The code blocks on the lower sequence beam

Finally, the code on the lower sequence beam of the main program continuously reads the *Total Balls* variable—which stores the number of balls that Guard-Bot has thrown—converts the number data into text data with a Number to Text block, uses a Text block to combine this text data with the text "Balls Thrown: " (notice the space after the colon), and finally displays it on the LCD with a Display block (Figure 15-14).

To test the Guard-Bot2 program, place Guard-Bot on the NXT test pad, and make sure there are several feet of empty space around the test pad. Next, synchronize the legs by using the knob wheels to adjust their positions; one set of legs should look like Figure 15-15, and the other set of legs should look like Figure 15-16. When you're finished, Guard-Bot should be standing on one leg on one side and two legs on the other side. You must ensure that the legs are synchronized every time you run the program. If you don't, the robot won't work properly!

NOTE If you did not properly position the gears when you were building the robot, you'll need to pull a gear out and then put it back in place in order to synchronize the legs as shown in Figures 15-15 and 15-16.

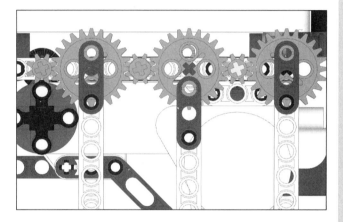

Figure 15-15: On one side (either left or right), you must position the legs as shown here.

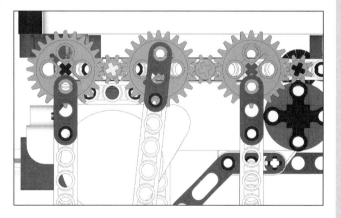

Figure 15-16: On the other side, you must position the legs as shown here.

Start the robot, step back, and watch for a moment. Two legs on each side should be synchronized with a third leg on the other side, so that Guard-Bot always has three legs on the ground. Move toward the front of Guard-Bot until the robot detects you (it should say, "Aha!") and throws its ball at you. After that, Guard-Bot should stop and wait until you bump its touch sensor, letting it know that you've reloaded its ball.

Next, let Guard-Bot walk to the black line on the test pad, and watch what happens. Guard-Bot should promptly take three steps backward and then change direction by running its legs in opposite directions. Now allow Guard-Bot to walk for six seconds without detecting the black line or someone in the room. When that happens, the robot should promptly change direction again and then continue walking forward. Note that if Guard-Bot detects someone while changing direction in either of these cases, it does immediately throw the ball but will not stop and wait for the ball to be reloaded until it has finished its move.

You might notice that Guard-Bot's legs sometimes get slightly out of sync for a moment. While the servo motors do effectively drive Guard-Bot, they don't perform perfectly. Nevertheless, for Guard-Bot's purposes, using the motors' built-in rotation sensors to synchronize the legs is still a practical solution.

NOTE Using fresh batteries helps Guard-Bot's legs to stay synchronized.

conclusion

Building a walking or legged NXT robot can be a challenging task, which Guard-Bot certainly demonstrates. In this chapter you began by building Guard-Bot and its eleven subassemblies, using quite a few pieces from the NXT set in the process. Then you programmed Guard-Bot in two different ways, first as a stationary motion detector and then as an intruder detector that walks around and guards the NXT test pad. And besides all this, you even learned how NXT robots can throw balls!

16

golf-bot

After creating a variety of mobile robots, in this chapter we'll finally build and program a stationary robot: Golf-Bot (Figure 16-1). This creation's goal is to hit a ball into a target made from pieces in the NXT set (the ball comes from the NXT set, as well). You can place the target anywhere within a range of two feet from the robot, and Golf-Bot will find the target by rotating on a turntable and searching the area with an ultrasonic sensor. Upon detecting the target, Golf-Bot opens a

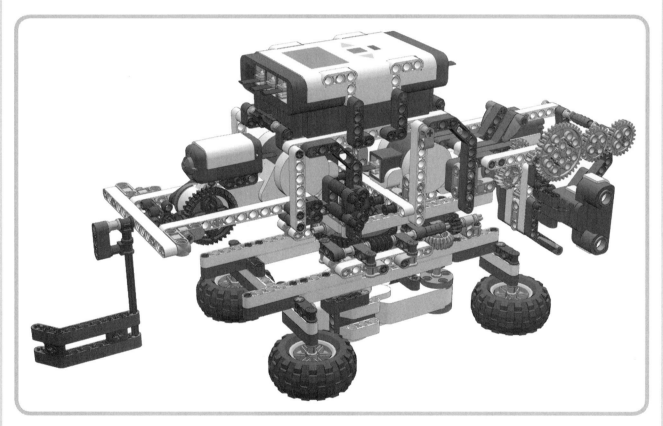

Figure 16-1: Golf-Bot is a stationary robot that hits a ball into a target.

hand that holds a ball and places the ball on the ground. The robot then turns around and hits the ball using a club-like structure on its other side. If the ball lands inside the target, you bump (i.e., press and release) a touch sensor on the robot to confirm the successful shot. If the ball lands too far to the right or left of the target, you can use the Left and Right buttons on the NXT to tell Golf-Bot how to adjust its shot. When you drop the ball back into its hand, a sound sensor detects the noise made by the falling ball, and the robot proceeds to place and then hit the ball again. Golf-Bot repeats this process with you until the ball lands inside the target and you then bump its touch sensor.

First, you'll build Golf-Bot, which is the largest and most complex robot in this book. Besides using three servo motors and three sensors, Golf-Bot employs three compound gear trains and all three types of "other" gears in the NXT set: the worm gear, the turntable, and the knob wheel. After construction, you'll program and test the robot, first creating an interactive program for operating the turntable (i.e., rotating the robot to a desired position) and then creating the main program for playing Golf-Bot's version of golf.

building golf-bot

Golf-Bot is composed of nine subassemblies. After you've built them all and combined them in the final assembly, you'll construct the target. Figure 16-2 shows the BOM for Golf-Bot (including the target)—it uses most of the pieces in the NXT set! You'll build the following subassemblies in the order shown:

* Turntable subassembly
* Turntable Driver subassembly
* Base subassembly
* Club subassembly
* Club Driver subassembly
* Ultrasonic Sensor subassembly
* Hand subassembly
* Middle Structure subassembly
* Top subassembly

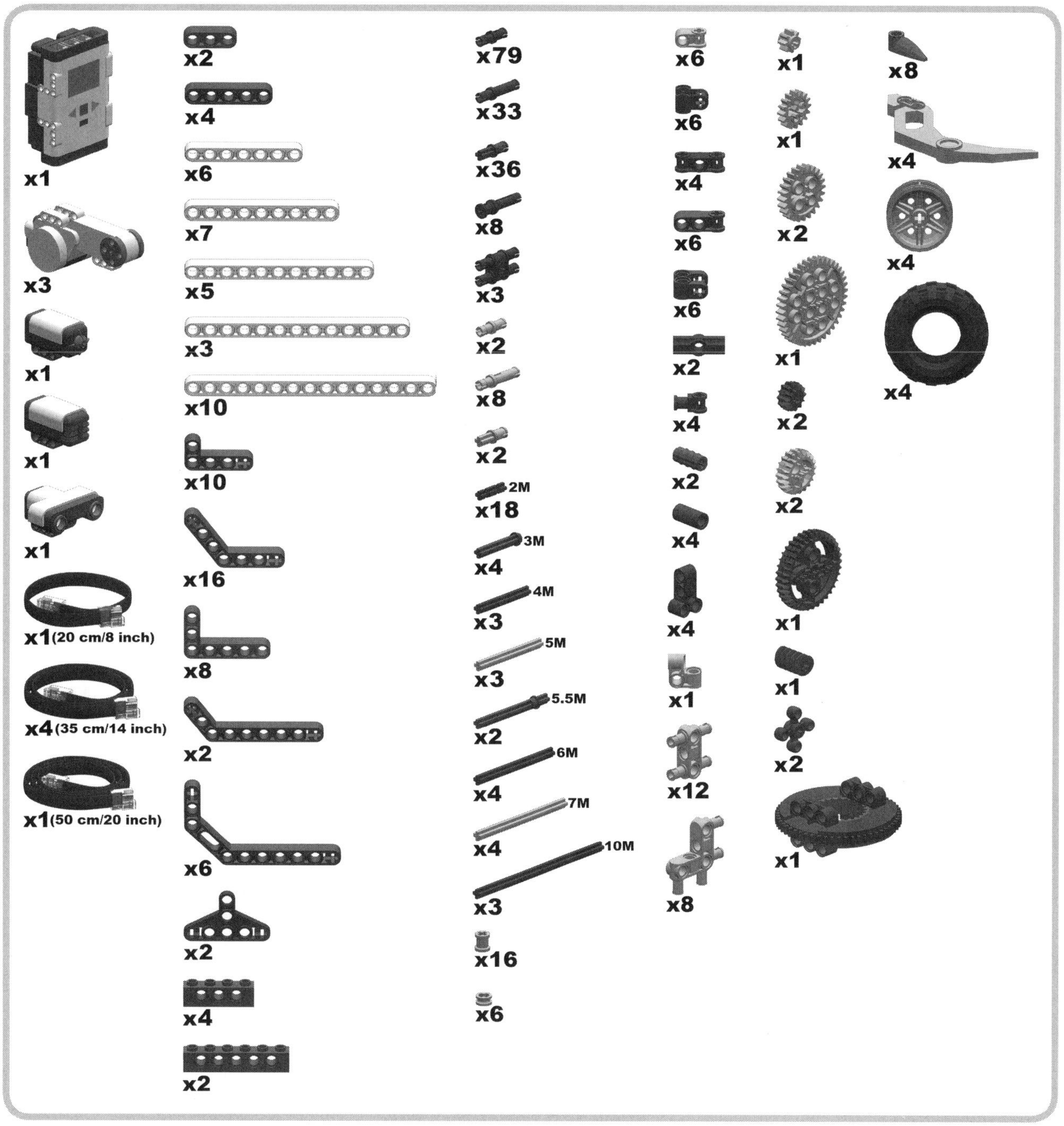

Figure 16-2: A BOM for Golf-Bot

turntable subassembly

We'll begin with the Turntable subassembly, which we'll attach to the Base subassembly later. The *Turntable subassembly* features a turntable, a worm gear, and significant bracing that helps to keep the two gears meshed. You'll connect the very large Top subassembly to the turntable in the final assembly, so the bracing is necessary to help prevent the gears from separating under the substantial weight. In addition, the gearing must provide enough torque to move the Top subassembly, so gearing down is important as well. A worm gear driving the turntable produces a gear ratio of 56:1, but you'll add some additional gearing in the Base subassembly that results in a final gear ratio of approximately 34:1 (which is still very powerful). There are eight construction steps.

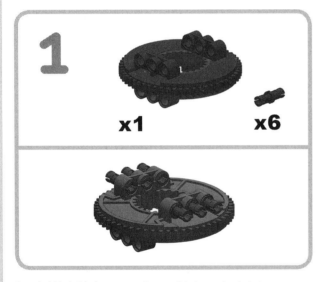

Step 1: Add six friction pegs to the round-holes on the dark stone gray part of the turntable.

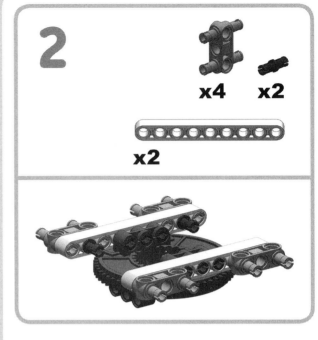

Step 2: Add two 9M beams, four 3M pegged blocks, and two friction pegs.

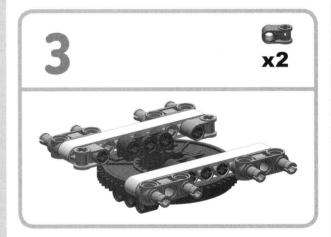

Step 3: Add two cross blocks.

4 x2 x2

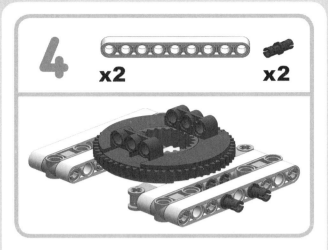

Step 4: Flip the model over so that the black part of the turntable is on the top and the cross blocks are facing you. Then add two 9M beams and two friction pegs.

5 x2 **2M** x2

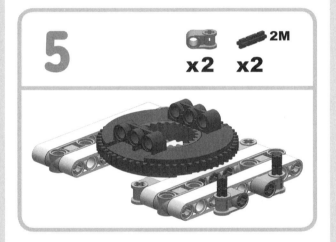

Step 5: Add two cross blocks and two 2M notched axles.

6 x1 x2 **10M** x1 x2 x2

Step 6: Add a worm gear, two half-bushings, two extended cross blocks, and two bushings to a 10M axle. Then push the extended cross blocks onto the 2M notched axles.

7 x2 **2M** x2

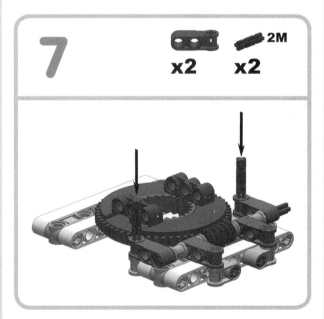

Step 7: Add two more extended cross blocks and two more 2M notched axles.

8 x2 x4

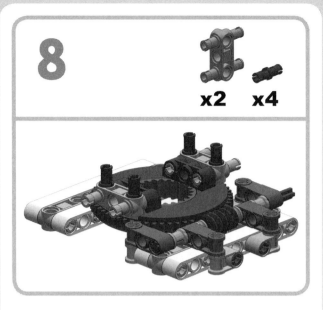

Step 8: Finish by adding two 3M pegged blocks and four friction pegs.

turntable driver subassembly

Now we'll build the *Turntable Driver subassembly*, which attaches to the Base subassembly that you'll build next and powers the Turntable subassembly. While the Turntable Driver subassembly consists of only a few pieces, assembling them separately in this subassembly makes attaching them to the Base subassembly much easier. There are three construction steps.

1 x1

7M

x1 x1

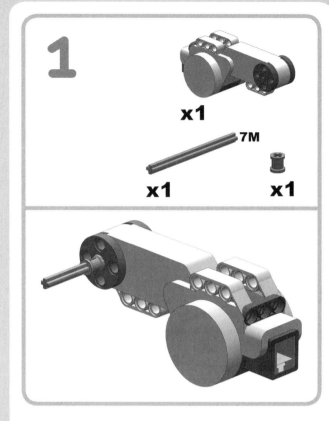

Step 1: Connect a 7M axle and a bushing to a motor.

2 x1 x1 x1

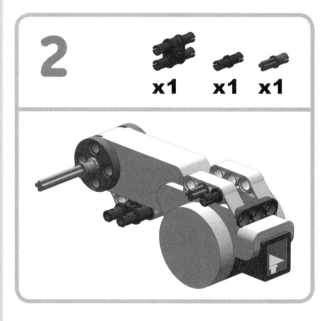

Step 2: Add a double friction peg, a friction peg, and a friction axle peg.

3

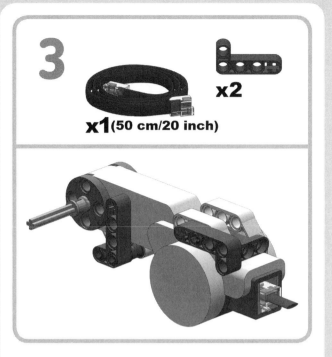

x1 (50 cm/20 inch) x2

Step 3: Finish by adding two 5M perpendicular angled beams and then connecting a 50 cm/20 inch cable to the motor. You'll connect this cable to the NXT in the final assembly.

base subassembly

The *Base subassembly* sits on four balloon wheels and tires (proof that tires and wheels can be used for creations other than vehicles!) and employs a large number of beams and pegs to ensure that the base is strong enough to support the rest of the robot. In addition, you'll attach the Turntable and Turntable Driver subassemblies here, and you'll even assemble a pair of double bevel gears on perpendicular axles. There are ten construction steps; pay careful attention to the positions of the numerous pegs.

1

x10 x4

x2

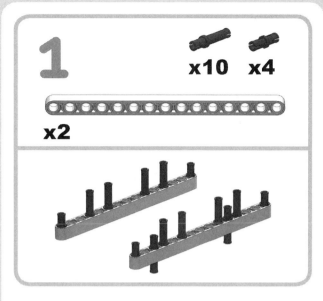

Step 1: Add ten 3M friction pegs and four friction pegs to two 15M beams.

2

x5 x4

x4

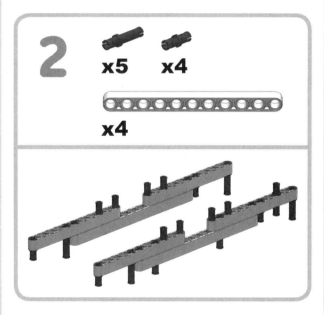

Step 2: Add four 11M beams, five more 3M friction pegs, and four more friction pegs.

3

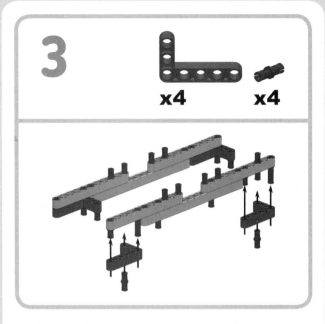

Step 3: Add four 7M perpendicular angled beams and four friction pegs, connecting one friction peg to each angled beam.

5

6M

x4

x4   x4

Step 5: Add four 6M axles, four inverted cross blocks, and four 5M beams.

4

x8   x4

x2

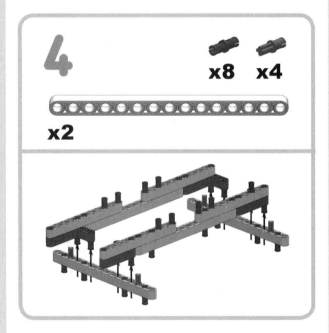

Step 4: Add two 15M beams, eight friction pegs, and four friction axle pegs. You should put the same types of pegs in the same round-holes on the back 15M beam and the front 15M beam.

6

x4   x4   x4

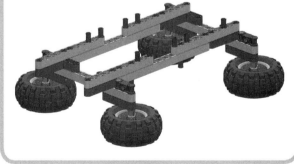

Step 6: Add four bushings to the 6M axles, and then attach four balloon wheels and four balloon tires to the 6M axles.

7

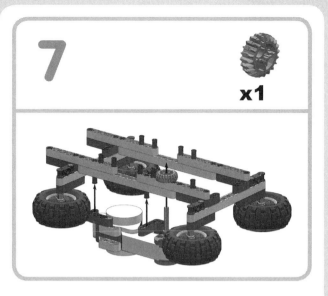

x1

Step 7: Connect the Turntable Driver subassembly you built earlier, and also attach a 20t double bevel gear to its 7M axle.

8

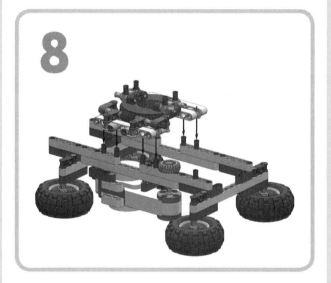

Step 8: Attach the Turntable subassembly you built earlier by snapping it into the friction pegs as shown.

9

x1   **x1**

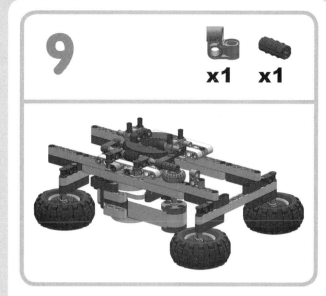

Step 9: Add a cornered peg joiner and an axle extender.

10

x2   **x1**

5M

x1

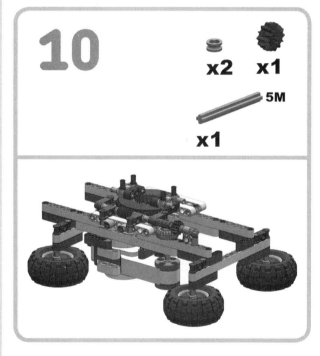

Step 10: Finish by adding a 5M axle, two half-bushings, and a 12t double bevel gear.

club subassembly

The *Club subassembly* attaches to the Club Driver subassembly that you'll build next, but the Club subassembly—and specifically its two 9M angled beams—is the part that actually hits the ball. There are three construction steps.

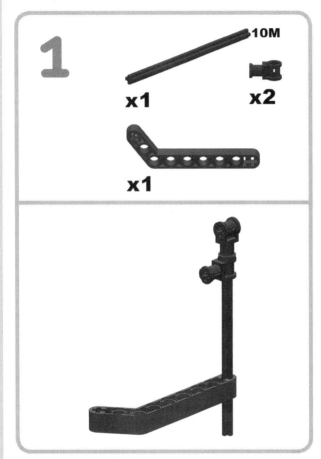

Step 1: Add two catches and a 9M angled beam to a 10M axle.

Step 2: Add two double friction pegs, a 5M axle, and a friction axle peg.

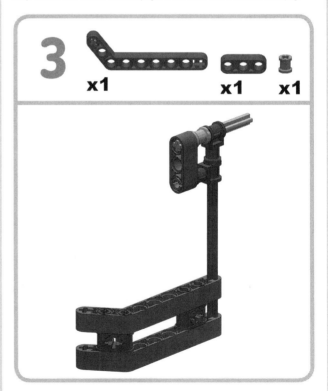

Step 3: Finish by adding another 9M angled beam, a bushing, and a 3M beam. The 3M beam serves as a brace that keeps the 10M axle in place.

club driver subassembly

The *Club Driver subassembly* powers the Club subassembly using a gear ratio of 1:3. Gearing up enables the Club sub-assembly to swing more rapidly and hit the ball farther. There are 15 construction steps.

Step 1: Add two 3M friction pegs, a friction peg, and a 3M pegged block to a servo motor.

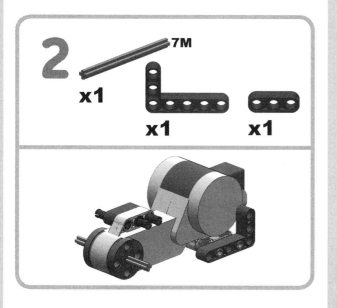

Step 2: Add a 7M perpendicular angled beam, a 3M beam, and a 7M axle.

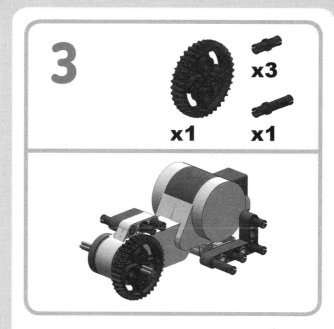

Step 3: Add a 36t double bevel gear to the 7M axle, and then add three friction pegs and a 3M friction peg to the 7M perpendicular angled beam.

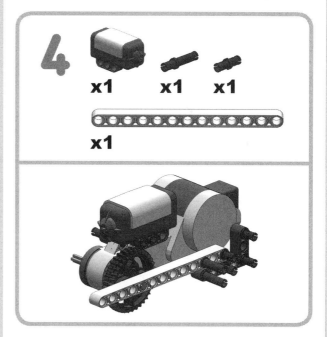

Step 4: Add a touch sensor and then a 13M beam with a friction peg and a 3M friction peg. You'll secure this beam in the next step.

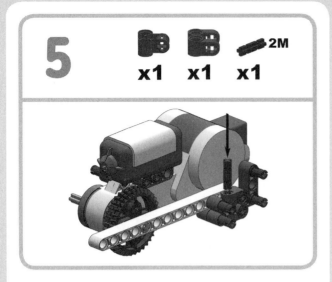

5 x1 x1 x1 **2M**

Step 5: Add a double cross block, a split cross block, and a 2M notched axle.

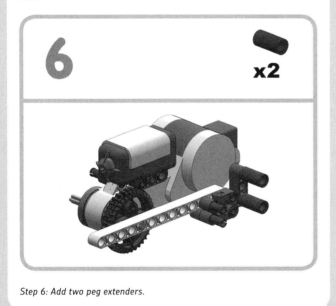

6 x2

Step 6: Add two peg extenders.

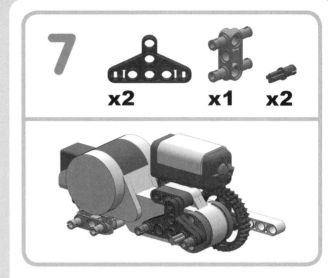

7 x2 x1 x2

Step 7: Flip the model over to the other side, and add two triangular half-beams, a 3M pegged block, and two friction axle pegs.

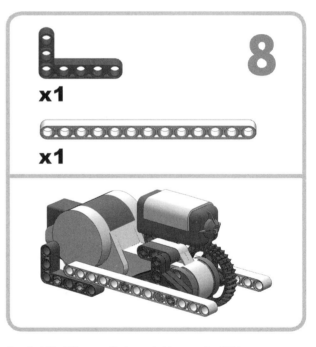

8 x1 x1

Step 8: Add a 7M perpendicular angled beam and a 13M beam.

9

x2 **x4**

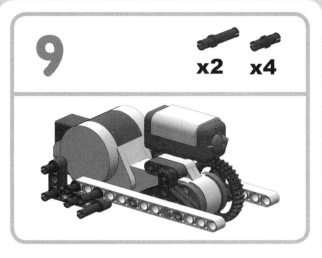

Step 9: Add two 3M friction pegs and four friction pegs.

11

x2

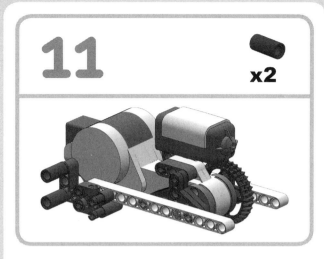

Step 11: Add two peg extenders.

10

x1 **x1** **2M** **x1**

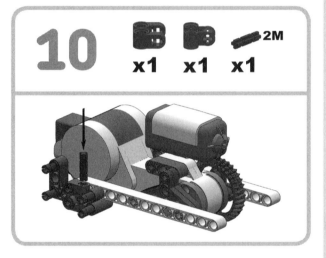

Step 10: Add a double cross block, a split cross block, and a 2M notched axle.

12

x2

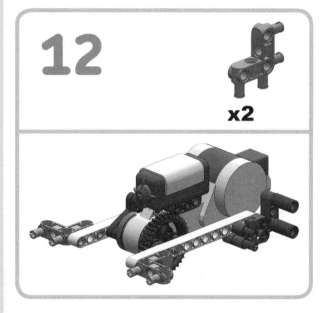

Step 12: Flip the model to the front, and add two 5M pegged perpendicular blocks.

13

x1 x1
10M
x1 x2

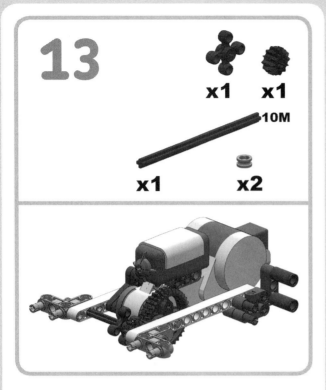

Step 13: Add a 10M axle with a 12t double bevel gear and a knob wheel. Add a half-bushing to each end of the axle.

14

x1
x1

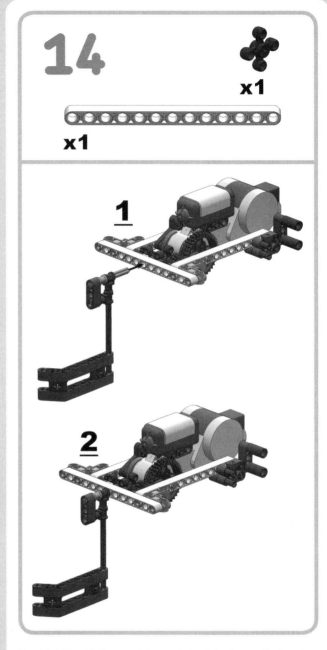

1

2

Step 14: Add a 13M beam and then push the Club subassembly through the beam and into a knob wheel. This second knob wheel should mesh with the first one.

15

x2 (35 cm/14 inch)

Step 15: Connect one 35 cm/14 inch cable to the motor and connect another 35 cm/14 inch cable to the touch sensor. You will connect these cables to the NXT when you build the Top subassembly.

ultrasonic sensor subassembly

The *Ultrasonic Sensor subassembly* attaches to the Hand subassembly that you'll build next. As discussed at the beginning of this chapter, the ultrasonic sensor detects the target. There are three construction steps.

1

x1   **x2**

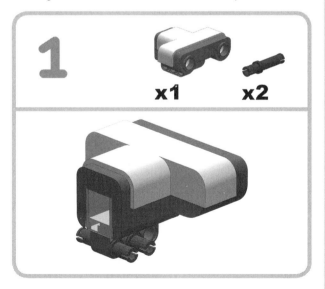

Step 1: Add two 3M friction pegs to the ultrasonic sensor.

2

x2   **2M x2**

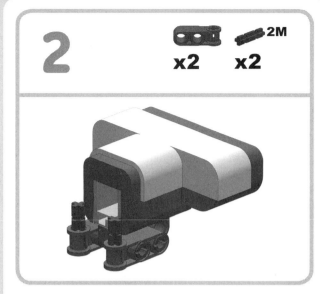

Step 2: Add two extended cross blocks and two 2M notched axles.

3

x1 (20 cm/8 inch)   **x2**

Step 3: Add two catches and connect a 20 cm/8 inch cable to the sensor. You will connect this cable to the NXT when you build the Top subassembly.

hand subassembly

The *Hand subassembly* uses a servo motor and an extensive gear train to open and close a uniquely designed hand. When the hand is in the closed position, you can drop a ball into it from above. A sound sensor detects the ball as it rattles into place, enabling the robot to know when a ball has been loaded into the hand. When the hand opens, the ball is guided onto the ground by surrounding pieces so that it consistently falls in the same spot relative to the Club subassembly. There are 22 construction steps.

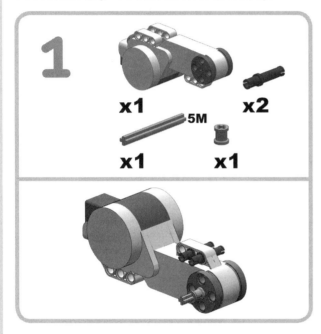

Step 1: Add a 5M axle, a bushing, and two 3M friction pegs to a servo motor.

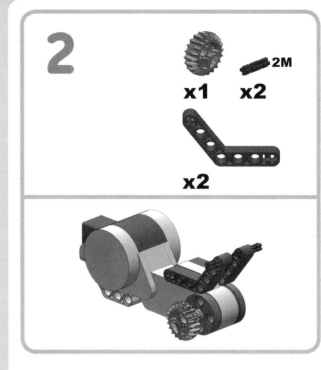

Step 2: Add two 7M angled beams, two 2M notched axles, and a 20t double bevel gear.

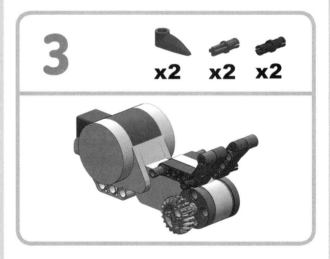

Step 3: Add two TECHNIC teeth, two friction axle pegs, and two friction pegs.

4

x1 7M

x1

x1

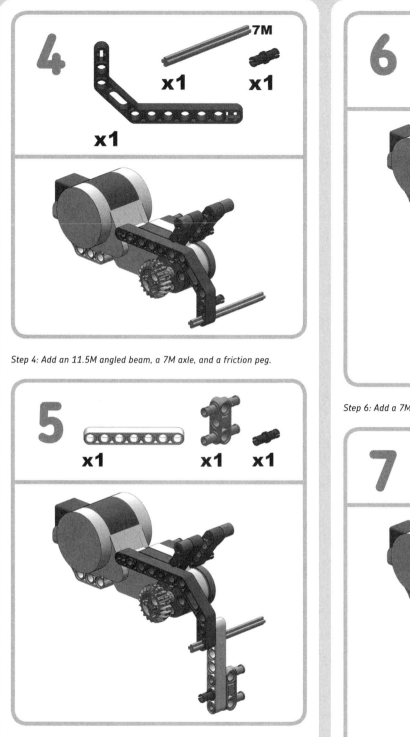

Step 4: Add an 11.5M angled beam, a 7M axle, and a friction peg.

5

x1 x1 x1

Step 5: Add a 7M beam, a 3M pegged block, and a friction peg.

6

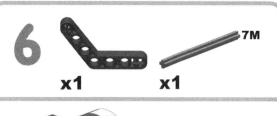

x1 7M

x1

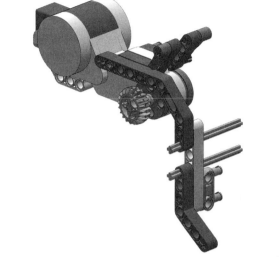

Step 6: Add a 7M angled beam and a 7M axle.

7

x1 x1

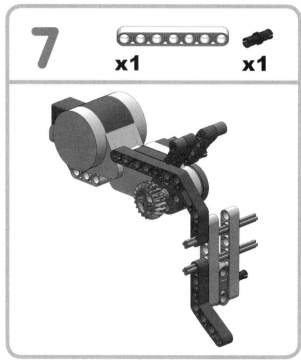

Step 7: Add another 7M beam and a friction peg.

8

x1   **2M** x2

x1

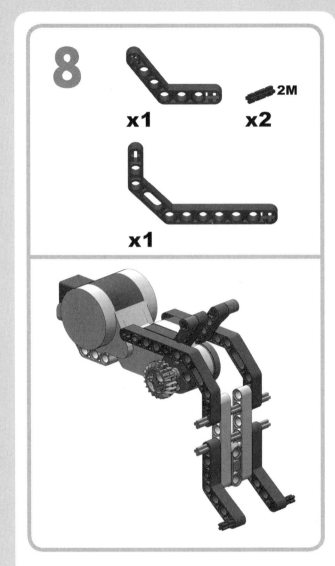

Step 8: Add an 11.5M angled beam, a 7M angled beam, and two 2M notched axles to the ends of the 7M angled beams.

9

x4

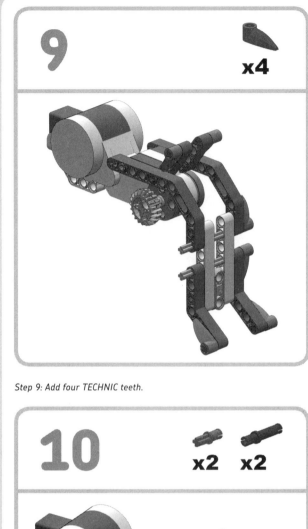

Step 9: Add four TECHNIC teeth.

10

x2   x2

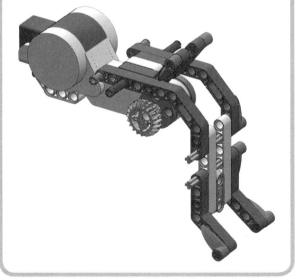

Step 10: Near the back of the subassembly, add two friction axle pegs and two 3M friction pegs to the 11.5M angled beams.

11

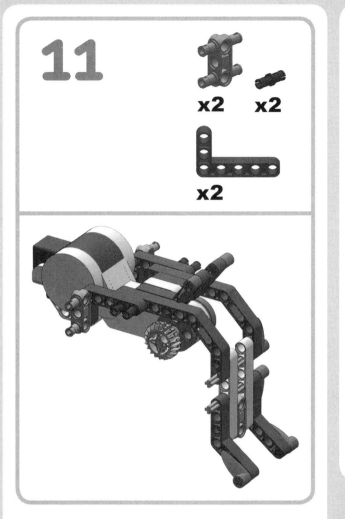

x2   x2

x2

Step 11: Add two 7M perpendicular angled beams, two 3M pegged blocks, and two friction pegs.

12

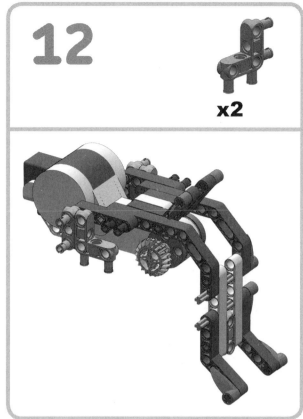

x2

Step 12: Add two 5M pegged perpendicular blocks.

13

4M

x1 x2

x1 x1

x1

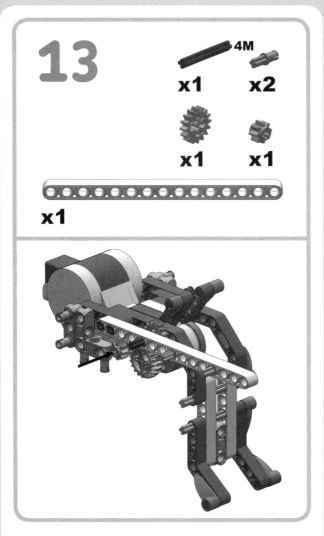

Step 13: Add a 15M beam and two axle pegs to the 15M beam. Also add a 4M axle, a 16t gear, and an 8t gear. The 16t gear should mesh with the 20t double bevel gear.

14

x1 x1

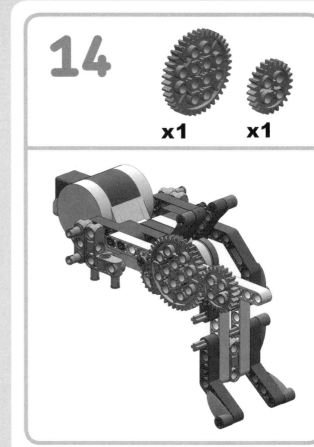

Step 14: Add a 40t gear and a 24t gear to the axle pegs.

15

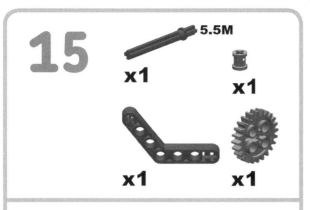

5.5M

x1

x1

x1

x1

Step 15: Add a bushing and a 7M angled beam to a 5.5M stopped axle. Then slide the axle through the 15M beam as shown, and attach a 24t gear to it.

16

5.5M

x1

x1

x1

x1

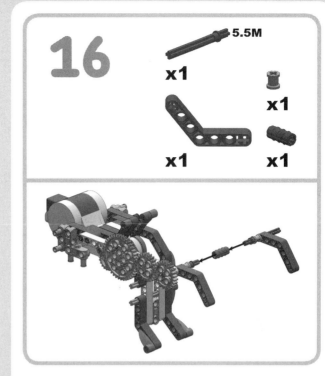

Step 16: Add an axle extender as well as another 5.5M stopped axle, a bushing, and a 7M angled beam.

17

x4   x1

x1

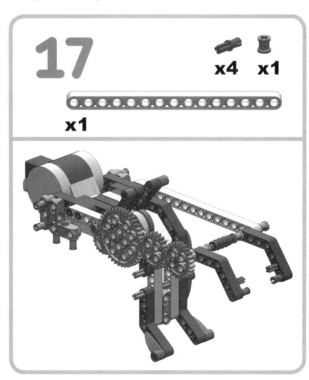

Step 17: Add a 15M beam and a bushing. Then attach four friction axle pegs to the ends of the 7M angled beams.

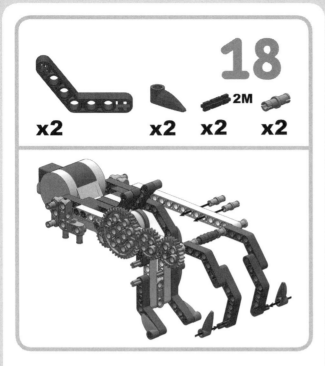

Step 18: Add two 7M angled beams, two 2M notched axles, two TECHNIC teeth, and two pegs.

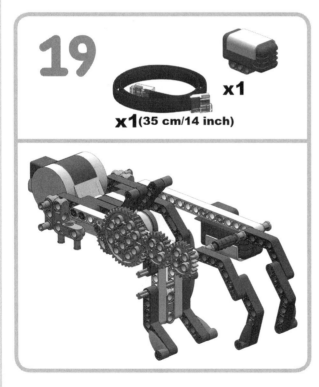

Step 19: Add the sound sensor to the pegs, and connect a 35 cm/14 inch cable to the sensor. You will connect this cable to the NXT when you build the Top subassembly.

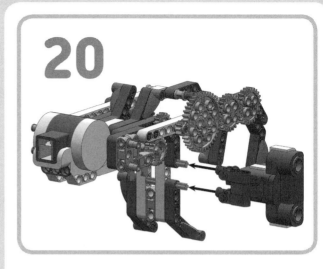

Step 20: Push the catches of the Ultrasonic Sensor subassembly you built earlier into the axles on the side of the Hand subassembly, as shown.

NOTE Make sure that you pass the ultrasonic sensor's cable in between the two catches so that the cable is directed away from the hand. You don't want the hand to get caught on the cable!

Step 21: Connect a 3M pegged block to the back of the motor.

22

x4

x1 (35 cm/14 inch)

x1

Step 22: Connect a 35 cm/14 inch cable to the motor, and connect an 11M beam with four bushed friction pegs to the 3M pegged block. When you build the Top subassembly, you will connect the cable to the NXT and also push the bushed friction pegs through the 11M beam and into the Top subassembly.

middle structure subassembly

The *Middle Structure subassembly* serves as a sort of central structure for attaching the NXT, Club Driver subassembly, and Hand subassembly to the robot. Specifically, when you build the Top subassembly, you'll firmly attach the Club Driver and Hand subassemblies to the Middle Structure subassembly. There are 10 construction steps.

1

x4

x4   x2

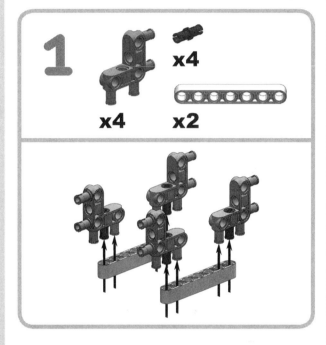

Step 1: Connect four friction pegs to four 5M pegged perpendicular blocks, and then attach two 7M beams to them.

2

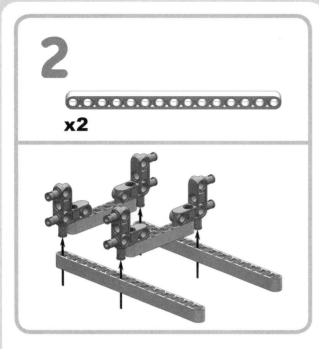

x2

Step 2: Add two 15M beams.

3

x2

x2

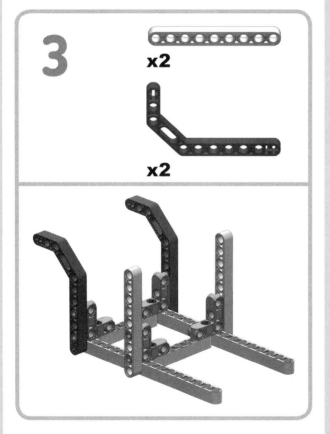

Step 3: Add two 9M beams and two 11.5M angled beams.

4

x2 **x2** **x4**

Step 4: Push two 3M friction pegs through the 11.5M angled beams and into the 5M pegged perpendicular blocks. Also add two friction axle pegs and four friction pegs to the 11.5M angled beams.

5

x2 **x2** **2M** **x2** **x4**

Step 5: Add two double cross blocks, two split cross blocks, two 2M notched axles, and four bushed friction pegs. When you're building the Top subassembly, you'll push the bushed friction pegs all the way through.

6

x4 x2 x2

x2

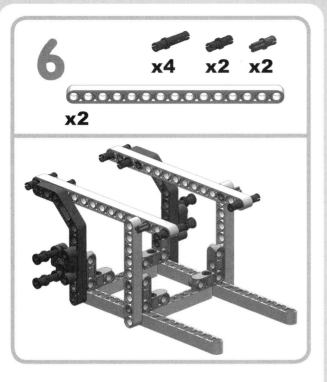

Step 6: Add two 15M beams, four 3M friction pegs, two friction pegs, and two friction axle pegs.

7

x4 4M x2 x2

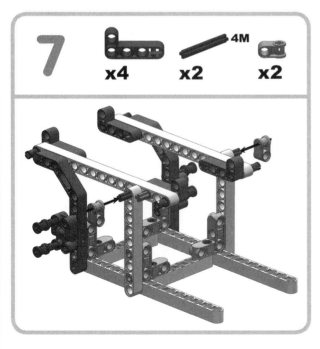

Step 7: Add four 5M perpendicular angled beams, two 4M axles, and two cross blocks. Because of the limited number of bushings in the NXT set, we use the cross blocks to hold the 4M axles in place.

8

x2 x4 3M x4

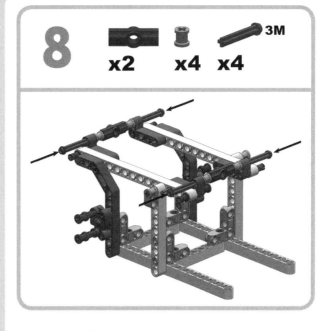

Step 8: Add two #2 angle connectors, four bushings, and four 3M studded axles. Together, these pieces firmly connect the two sides of the subassembly.

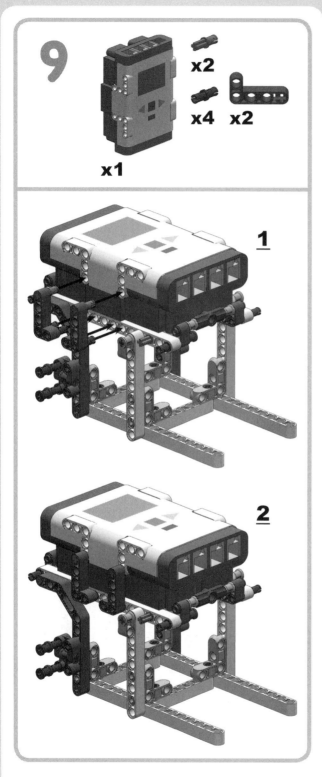

Step 9: Place the NXT on the top of the subassembly, and secure one side with two 5M perpendicular angled beams, two friction axle pegs, and four friction pegs.

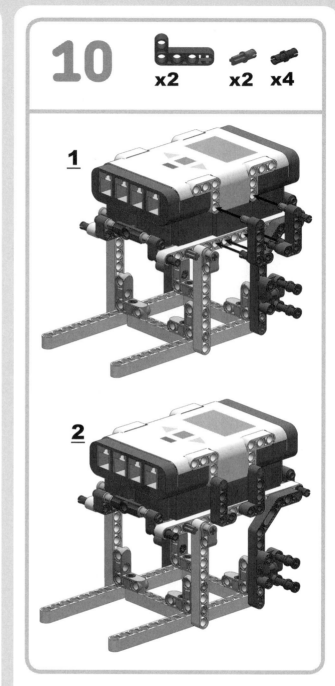

Step 10: Secure the other side with two 5M perpendicular angled beams, two friction axle pegs, and four friction pegs.

top subassembly

The *Top subassembly* combines the Middle Structure, Hand, and Club Driver subassemblies. In the final assembly, you'll snap the Top subassembly into the friction pegs on the turntable. There are five construction steps.

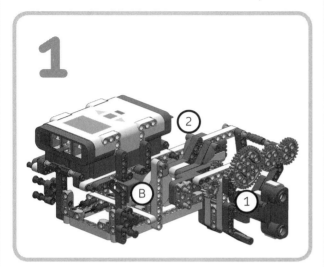

Step 1: Connect the Hand subassembly to the Middle Structure subassembly, snapping its 5M pegged perpendicular blocks into the 15M beams and pushing its bushed friction pegs into the 5M pegged perpendicular blocks on the Middle Structure subassembly. Connect the sensors and motor to the ports shown.

NOTE You'll need to tilt the Hand subassembly in order to get the 11M beam on the back of the motor past the 9M beams on the Middle Structure subassembly.

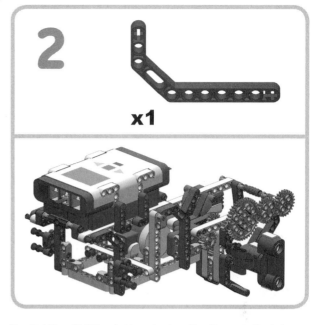

Step 2: Add an 11.5M angled beam to strengthen the connection between the Hand subassembly and the Middle Structure subassembly.

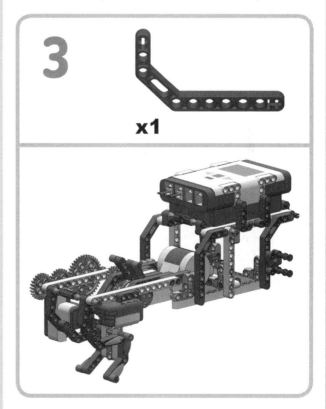

Step 3: Flip the model around to the other side and add another 11.5M angled beam.

Step 4: Flip the model back around to the original view, and add the Club Driver subassembly. To do this, push the bushed friction pegs on the Middle Structure subassembly into the peg extenders. Connect the touch sensor to port 3 and the motor to port C on the NXT.

x2

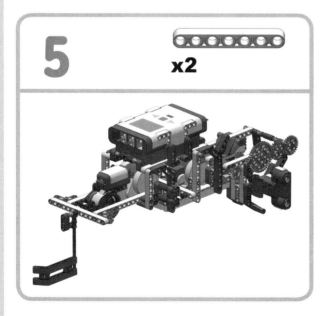

Step 5: Add two 7M beams to strengthen the connection between the Club Driver subassembly and the Middle Structure subassembly.

final assembly

The final assembly has only *two* construction steps—you've already done all the actual building! You just need to connect the Top subassembly to the Base subassembly and connect the motor in the Base subassembly to the NXT.

Step 1: Snap the Top subassembly onto the Base subassembly. The friction pegs should connect to two 7M beams on the Top subassembly.

2

Step 2: Connect the motor in the Base subassembly to port A. To do this, first take the cable and pass it through the hole in the middle of the turntable. Once you've done that, you can pull the cable out through the side of the Top subassembly and then connect it to the NXT.

NOTE Make sure that you pass the motor's cable through the hole in the turntable as described in step 2. Doing this takes a little maneuvering with your fingers, but it's crucial to the robot's performance!

the target

Before programming Golf-Bot, you need to build the *target*, which features four TECHNIC pincers that form an opening for the ball. Since the ultrasonic sensor must be able to detect the target, the target also has a structure that extends above the pincers. There are 11 construction steps.

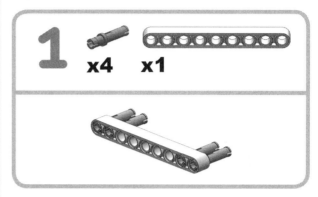

Step 1: Add four 3M pegs to a 9M beam.

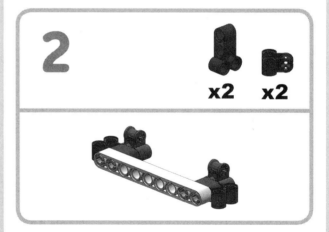

Step 2: Add two double peg joiners and two double cross blocks.

3

x2 **x2** 2M

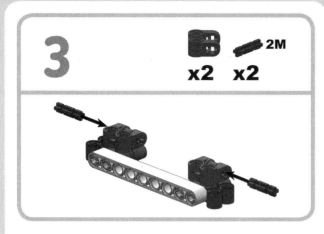

Step 3: Add two split cross blocks and two 2M notched axles.

4

x4 **x4**

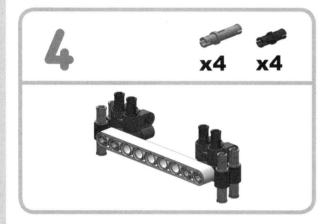

Step 4: Add four 3M pegs and four friction pegs.

5

x4

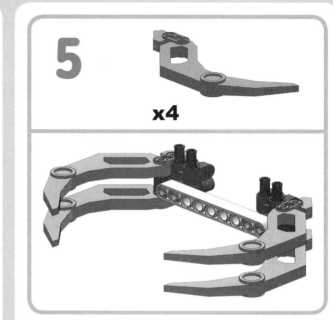

Step 5: Add four TECHNIC pincers.

6

x2 **x4**

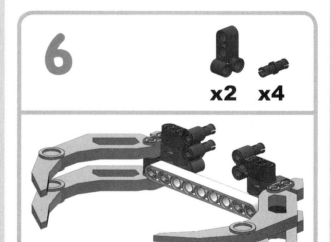

Step 6: Add two double peg joiners and four friction pegs.

7

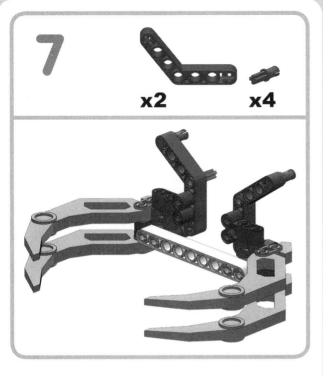

x2   x4

Step 7: Add two 7M angled beams and four friction axle pegs.

8

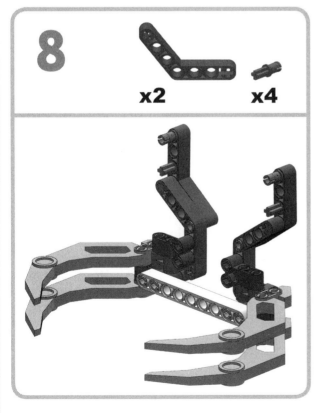

x2   x4

Step 8: Add two more 7M angled beams and four more friction axle pegs.

9

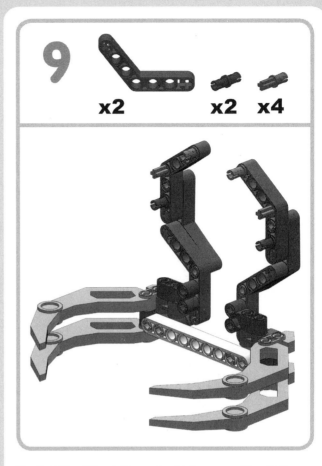

x2   x2   x4

Step 9: Add two 7M angled beams, two friction pegs, and four friction axle pegs.

10

x4

x2

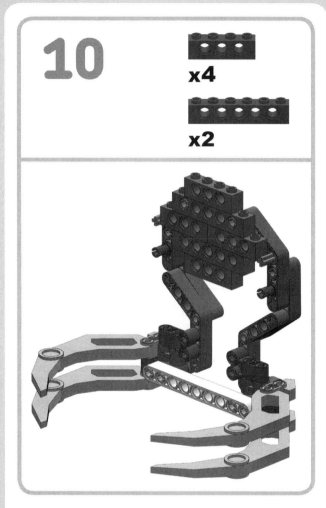

Step 10: Add four 1 × 4 TECHNIC bricks and two 1 × 6 TECHNIC bricks. (It's easiest if you connect the TECHNIC bricks to each other first, and then attach them to the model.)

11

x2

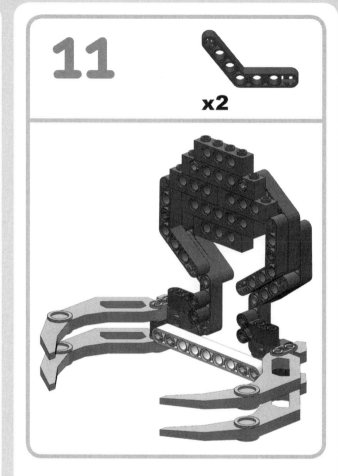

Step 11: Finally, add two more 7M angled beams.

programming and testing golf-bot

Programming Golf-Bot to find the target and hit a ball into it is not particularly difficult, but there is one potential problem that we must consider. Notice that all of the electronic pieces rotate *with* the base except for the motor that powers the base. The cable running from that motor to the NXT places a limitation on Golf-Bot: The robot cannot rotate forever in either direction because the cable would eventually get tangled up. Therefore, we should start Golf-Bot in the same position every time so that the robot can operate without encountering any cable trouble. To make this task easier, we'll create a program that you can use to power the turntable and quickly change Golf-Bot's position. After that, we'll create a separate program for Golf-Bot's main task—hitting a ball into the target.

positioning golf-bot: the golf-bot1 program

Our first program, Golf-Bot1 (Figure 16-3), is a relatively simple one that focuses on operating the turntable. An interesting aspect of this program, however, is that it utilizes three of the NXT's built-in sensors: the Left, Right, and Enter buttons. Let's break this program down into two sections and examine each one.

At the beginning of the program, three Display blocks give some instructions on how to use the program, and then a Wait block causes the program to wait one second before continuing (Figure 16-4). The three Display blocks put text on the LCD indicating that pressing the Right button instructs the robot to turn right, pressing the Left button instructs the robot to turn left, and pressing the Enter button stops the robot (but not the program).

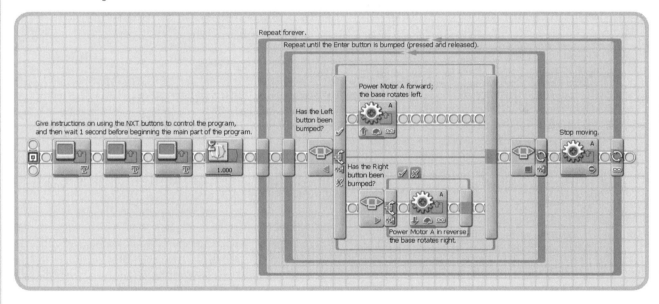

Figure 16-3: The Golf-Bot1 program

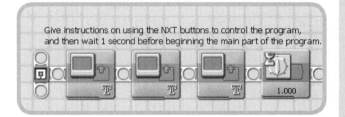

Figure 16-4: These four blocks start the Golf-Bot1 program.

The rest of the program accomplishes the real work (Figure 16-5). Inside a Loop block that repeats forever, another Loop block repeats until the Enter button is bumped (pressed and released). Inside this second Loop block, a Switch block asks, *Has the Left button been bumped?* If it has, Motor A begins driving forward, which rotates the robot to the left. If the Left button hasn't been bumped, another Switch block asks, *Has the Right button been bumped?* If it has, the Motor A begins driving in reverse, which rotates the robot to the right. If the Right button hasn't been bumped, the robot doesn't do anything. Finally, when you bump the

Enter button, a Move block stops Motor A. Pressing the Left or Right buttons will, of course, make the robot start rotating again.

Download the Golf-Bot1 program to Golf-Bot, and then run the program. The text explaining how to use the program should immediately display on the LCD and remain until you stop the program. To get the robot moving, press and release either the Left or the Right button on the NXT. Golf-Bot should immediately turn in the appropriate direction. Notice that if you bump the Right button while the robot is rotating left or if you bump the Left button while

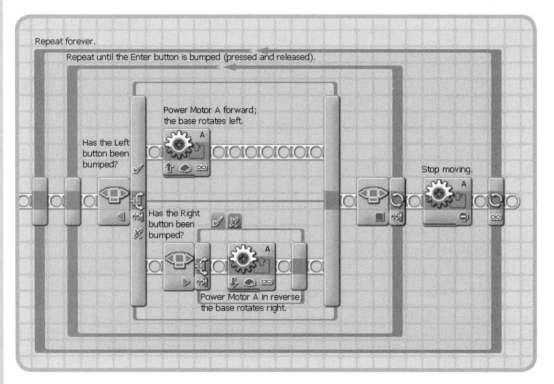

Figure 16-5: The main part of the Golf-Bot1 program uses the NXT buttons to control the motor that drives the turntable.

the robot is rotating right, Golf-Bot immediately begins rotating in the new direction. If you bump the Enter button while the robot is rotating, Golf-Bot should immediately stop.

Now you should practice positioning the robot for the next program. Although the next program does reposition Golf-Bot when it's finished, there will be times when you need to position Golf-Bot yourself by using the Golf-Bot1 program. Using the buttons on the NXT, position Golf-Bot so that it looks like Figure 16-1 on page 221. The Hand subassembly should be on the robot's left side, and the Club Driver subassembly should be on the robot's right side. Furthermore, if you had rotated the robot right from its original position, you should then rotate it left to return to the original position; and if you had rotated the robot left from its original position, you should then rotate it right to return to the original position. In this way, you can prevent the cable from getting tangled when Golf-Bot runs the main program (which we'll examine next).

NOTE You can also use the Golf-Bot1 program to test for any other cable tangles. Specifically, by rotating Golf-Bot right and left, you can determine whether the sensor or motor cables get caught on various places of the robot.

playing golf: the golf-bot2 program

Now let's move on to the program that Golf-Bot uses to find the target and hit a ball into it. The complete program, Golf-Bot2, is shown in Figure 16-6. Let's look at the individual sections of this program first and then test it.

NOTE The Golf-Bot2 program includes one My Block, the Hit Ball block. In order for the program to successfully load the block, the block must be in the same location on your computer as the Golf-Bot2 program (i.e., the program loads the block from its own directory). You can, however, add a copy of this block to the Custom palette if you want to use it in your own programs.

The first series of blocks in the program instruct the robot to wait for the ball to be loaded and then to start searching for the target (Figure 16-7). First, a Wait block waits until it detects a sound greater than 100 percent (i.e., the ball being dropped into the Hand subassembly). When that condition has been met, a Sound block plays a tone to let you know that the ball has been detected. Next, inside a Loop block that repeats until it receives *true* logic data, a Motor block rotates Golf-Bot, while a Logic block performs an *Or* operation on data from an Ultrasonic Sensor block and

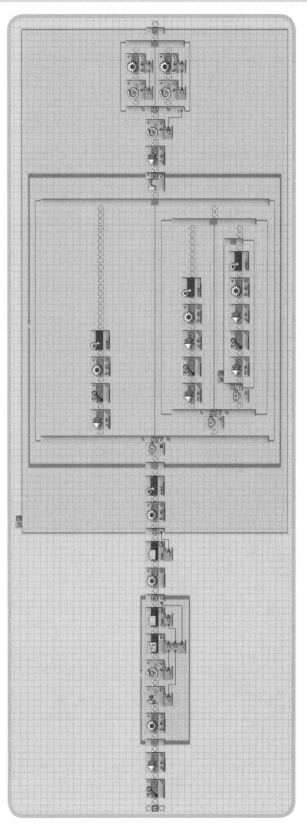

Figure 16-6: The Golf-Bot2 program (shown in detail in Figures 16-7 through 16-13)

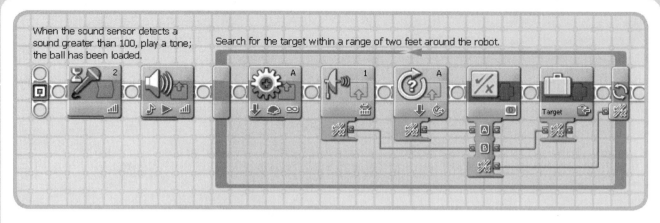

Figure 16-7: The first series of blocks in the Golf-Bot2 program

a Rotation Sensor block. When the ultrasonic sensor detects an object less than two feet away or when the motor in port A has run for 34 rotations (about one complete rotation for Golf-Bot), the Logic block sends a *true* value to the Loop block. Also, if the ultrasonic sensor does detect the target, the Target variable—which stores logic data—becomes *true*.

After the Loop block has finished, a Motor block immediately stops the motor rotating the robot, and the Target variable sends its data to a Switch block. If the logic data is *false* (Golf-Bot didn't detect a target), another Motor

block immediately rotates Golf-Bot back to the starting position, a Sound block plays a tone, and the program ends (Figure 16-8). However, if the logic data is *true* (Golf-Bot did detect the target), a Motor block rotates Golf-Bot for an additional 0.5 rotations and then the program executes a My Block (Figure 16-9). The My Block is called *Hit Ball* (Figure 16-10), and it consists of a series of Motor blocks that carefully place the ball on the ground, turn the robot around, and hit the ball with the Club subassembly.

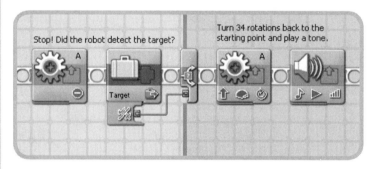

Figure 16-8: If the Target variable sends false logic data to the Switch block, Golf-Bot rotates back to its starting position and plays a tone. The program then ends.

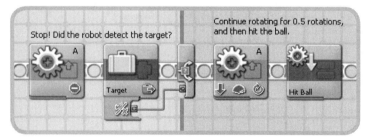

Figure 16-9: If the Target variable sends true logic data to the Switch block, Golf-Bot continues rotating for 0.5 rotations and then executes the Hit Ball My Block.

Once the robot has hit the ball, the program enters a Loop block that repeats until you bump the touch sensor, telling the robot that the ball is in the target. Inside this Loop block are three Switch blocks that ask, in turn, whether the Enter button, Left Button, and Right button have been bumped. First, a Switch block asks if you bumped the Enter button, which means you want Golf-Bot to repeat the same shot it just made (Figure 16-11). If you have bumped the Enter button, Golf-Bot plays a tone and waits for you to reload the ball, and then it rotates to the exact spot where it had previously put the ball and executes the Hit Ball My Block.

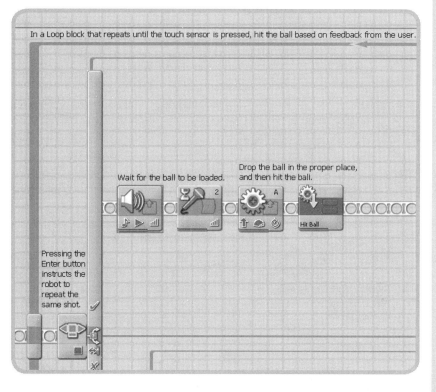

In a Loop block that repeats until the touch sensor is pressed, hit the ball based on feedback from the user.

Wait for the ball to be loaded.

Drop the ball in the proper place, and then hit the ball.

Pressing the Enter button instructs the robot to repeat the same shot.

Hit Ball

Figure 16-11: If you press the Enter button, this Switch block instructs Golf-Bot to repeat the same shot.

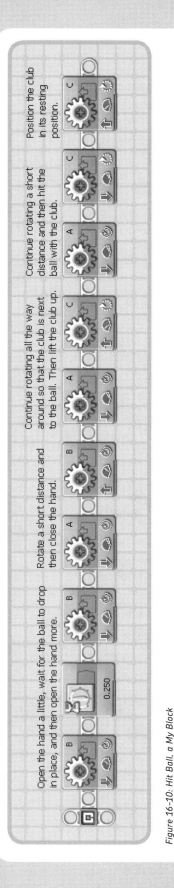

Position the club in its resting position.

Continue rotating a short distance and then hit the ball with the club.

Continue rotating all the way around so that the club is next to the ball. Then lift the club up.

Rotate a short distance and then close the hand.

Open the hand a little, wait for the ball to drop in place, and then open the hand more.

Figure 16-10: Hit Ball, a My Block

If the Enter button has not been bumped, the program enters another Switch block that asks if you bumped the Left button—meaning you want Golf-Bot to aim more to the left (Figure 16-12). If you have bumped the Left button, the Switch block's top sequence beam executes a Sound block that instructs Golf-Bot to say, "Left." After waiting for you to load the ball, Golf-Bot positions its hand slightly to the left of the previous spot where it placed the ball. Finally, it executes the familiar Hit Ball block to hit the ball. If you did not bump the Left button, another Switch block asks if you bumped the Right button, which means you want Golf-Bot to aim more to the right. If this is the case, Golf-Bot says, "Right," waits for you to reload the ball, and then positions its hand slightly to the right of the previous spot where it placed the ball. Of course, the program then executes the Hit Ball block. If none of the buttons have been bumped, nothing happens and the robot sits still.

Finally, when the ball has successfully gone into the target and you have bumped the touch sensor, a Sound block plays a triumphant tone. In addition, the robot returns to its original position by reading the built-in rotation sensor for the motor in port A to determine how far it should rotate (Figure 16-13).

Download the Golf-Bot2 program to Golf-Bot and place the robot in the middle of a carpeted room. (You cannot use Golf-Bot on a hard floor because the plastic balls from the NXT set will roll around too much.) First, make sure that there are several feet of empty space around Golf-Bot in all directions so that the robot doesn't accidentally detect an object other than the target. Second, make sure that Golf-Bot is in the proper starting position, using the Golf-Bot1 program as necessary. Third, place the target facing toward Golf-Bot and not farther than two feet from it. (If the target isn't facing the robot, Golf-Bot can't hit the ball into the target!) Fourth, make sure that the hand in the Hand subassembly is shut tight. Fifth, make sure that the club is in the downward or resting position (as shown in Figure 16-1).

Now run the Golf-Bot2 program, drop one of the balls from the NXT set into the Hand subassembly, and step back. Golf-Bot should play a tone, confirming that you've loaded the ball. Golf-Bot should then begin searching for the target by making a complete turn around its base.

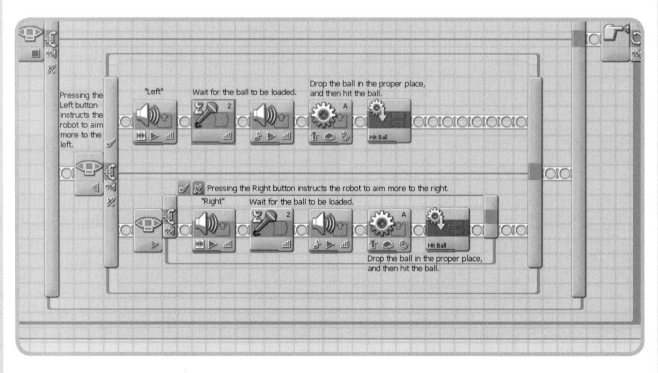

Figure 16-12: Pressing the Left or Right buttons instructs Golf-Bot to shoot the ball more to the left or right, respectively.

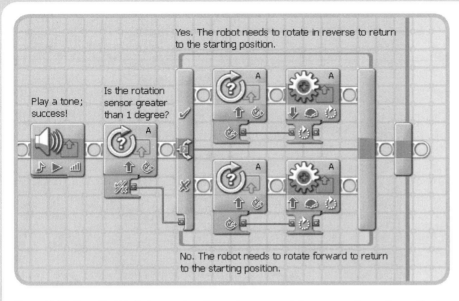

Yes. The robot needs to rotate in reverse to return to the starting position.

Play a tone; success!

Is the rotation sensor greater than 1 degree?

No. The robot needs to rotate forward to return to the starting position.

Figure 16-13: These final blocks instruct the robot to play a tone and then return to its starting position by reading the value of the built-in rotation sensor for Motor A.

NOTE If you placed the target too far away from Golf-Bot and the ultrasonic sensor doesn't detect it, simply reposition the target, and run the program again. Remember that if Golf-Bot doesn't detect anything, it will return to its starting position, and its program will end.

When Golf-Bot detects the target, it should immediately drop the ball onto the ground and continue rotating until its club is directly behind the ball. At that point, the robot should swing the club and knock the ball toward the target. If the ball lands inside the target, you can press the touch sensor. If the ball doesn't land inside the target, press the appropriate button on the NXT to adjust how the robot hits the ball, and then reload the ball. Remember that you can use the Enter button to instruct Golf-Bot to repeat a shot; this can be useful if pressing the Left and Right buttons seem to make Golf-Bot move too far to the left or right of the target.

Try placing the target in a variety of places around Golf-Bot and seeing how well the robot can detect the target and hit the ball into it. Sometimes Golf-Bot can get a ball into the target on the first try (that would be a hole in one for Golf-Bot!); other times Golf-Bot may need a number of adjustments to get the ball in the target. I also encourage you to adjust the settings of the blocks in the Golf-Bot2 program to see what happens. Can you also enhance the program by adding additional features?

conclusion

While mobile robots often seem to be the most common type of NXT inventions, stationary robots are also incredibly fun to build and program. In this chapter you created the stationary robot Golf-Bot, which can find a target, place a ball on the ground, and hit the ball into the target. You built nine subassemblies—some of which were quite complex—and combined them in a final assembly to construct the complete Golf-Bot. You then programmed Golf-Bot with two very different programs. The first program enabled you to position the robot by operating its turntable, and the second program instructed Golf-Bot how to play its version of golf.

You've finally come to the end of this book. Along the way, we've discussed a variety of topics, and I hope that you will pursue LEGO MINDSTORMS NXT further in any or all of the areas we've covered. There are always robots to invent, building techniques to discover, programming languages to learn, and new ideas and tools to try out. And as you create your very own NXT robots, don't be afraid to experiment with ideas that seem crazy—I've found that the crazy ideas are sometimes the ones that work the best. So start inventing and sharing your creations with the rest of the world. We want to see what you're creating, too!

LEGO MINDSTORMS NXT
piece library

Gaining a thorough understanding of the pieces in the NXT set is essential to becoming a proficient builder. Although Part II discusses the pieces in detail, this appendix gathers the basic facts about each piece for easy reference. In addition, simply reading the following material will help you quickly achieve familiarity with the more than 80 types of pieces. Table A-1 lists the categories and subcategories of pieces in the NXT set as presented in Chapter 4. Tables A-2 through A-6 correspond to the five main categories and give the name, color, quantity, and classification of those categories' respective pieces along with images of the pieces.

NOTE As mentioned in Chapter 4, LEGO sets usually include several extra of some of the smaller pieces. For this reason, your NXT set probably has a few more of some pieces than the official quantities given here.

table A-1: categories and subcategories

category	subcategories
Electronics	Microcomputers Motors Sensors Electrical cables
Beams	Straight beams Angled beams Half-beams TECHNIC bricks
Connectors	Pegs Axles Connector blocks
Gears	Spur gears Double bevel gears Other gears
Miscellaneous elements	n/a

table A-2: electronics

piece	color	quantity	classification
NXT	Mainly light stone gray and dark stone gray	1	Electronics → Microcomputers
Servo motor	Light stone gray with orange shaft heads	3	Electronics → Motors
Touch sensor	Dark stone gray and light stone gray with an orange push button	1	Electronics → Sensors
Light sensor	Mainly dark stone gray and light stone gray	1	Electronics → Sensors

table A-2: electronics (continued)

piece	color	quantity	classification
Sound sensor	Mainly dark stone gray and light stone gray	1	Electronics → Sensors
Ultrasonic sensor	Mainly dark stone gray and light stone gray	1	Electronics → Sensors
20 cm/8 inch electrical NXT cable	Black with transparent connectors	1	Electronics → Electrical cables
35 cm/14 inch electrical NXT cable	Black with transparent connectors	4	Electronics → Electrical cables
50 cm/20 inch electrical NXT cable	Black with transparent connectors	2	Electronics → Electrical cables

table A-3: beams

piece	color	quantity	classification
3M beam	Dark stone gray	16	Beams → Straight beams
5M beam	Dark stone gray	5	Beams → Straight beams
7M beam	Light stone gray	6	Beams → Straight beams
9M beam	Light stone gray	7	Beams → Straight beams
11M beam	Light stone gray	7	Beams → Straight beams
13M beam	Light stone gray	4	Beams → Straight beams
15M beam	Light stone gray	11	Beams → Straight beams
7M angled beam	Dark stone gray	16	Beams → Angled beams

table A-3: beams (continued)

piece	color	quantity	classification
9M angled beam	Dark stone gray	2	Beams → Angled beams
11.5M angled beam	Dark stone gray	6	Beams → Angled beams
5M perpendicular angled beam	Dark stone gray	10	Beams → Angled beams → Perpendicular angled beams
7M perpendicular angled beam	Dark stone gray	8	Beams → Angled beams → Perpendicular angled beams
Triangular half-beam	Black	4	Beams → Half-beams
1 × 4 TECHNIC brick	Dark stone gray	4	Beams → TECHNIC bricks
1 × 6 TECHNIC brick	Dark stone gray	2	Beams → TECHNIC bricks

table A-4: connectors

piece	color	quantity	classification
Peg	Medium stone gray	2	Connectors → Pegs
3M peg	Medium stone gray	13	Connectors → Pegs
Axle peg	Tan	4	Connectors → Pegs
Axle ball peg	Medium stone gray	2	Connectors → Pegs
Friction peg	Black	80	Connectors → Pegs → Friction pegs
3M friction peg	Black	34	Connectors → Pegs → Friction pegs
Friction axle peg	Blue	42	Connectors → Pegs → Friction pegs
Friction ball peg	Black	8	Connectors → Pegs → Friction pegs
Bushed friction peg	Black	8	Connectors → Pegs → Friction pegs
Double friction peg	Black	3	Connectors → Pegs → Friction pegs

table A-4: connectors (continued)

piece	color	quantity	classification
Steering link	Dark stone gray	3	Connectors → Pegs → Peg accessories
9M steering link	Black	2	Connectors → Pegs → Peg accessories
2M notched axle	Black	22	Connectors → Axles
3M axle	Medium stone gray	17	Connectors → Axles
3M studded axle	Dark stone gray	4	Connectors → Axles
4M axle	Black	4	Connectors → Axles
5M axle	Medium stone gray	7	Connectors → Axles
5.5M stopped axle	Dark stone gray	2	Connectors → Axles
6M axle	Black	4	Connectors → Axles

(continued)

table A-4: connectors (continued)

piece	color	quantity	classification
7M axle	Medium stone gray	4	Connectors → Axles
8M axle	Black	2	Connectors → Axles
10M axle	Black	4	Connectors → Axles
12M axle	Black	2	Connectors → Axles
Bushing	Medium stone gray	16	Connectors → Axles → Axle accessories
Half-bushing	Medium stone gray	6	Connectors → Axles → Axle accessories
Cross block	Medium stone gray	8	Connectors → Connector blocks

table A-4: connectors (continued)

piece	color	quantity	classification
Double cross block	Dark stone gray	6	Connectors → Connector blocks
Inverted cross block	Black	4	Connectors → Connector blocks
Extended cross block	Dark stone gray	6	Connectors → Connector blocks
Split cross block	Dark stone gray	6	Connectors → Connector blocks
#1 angle connector	Black	4	Connectors → Connector blocks
#2 angle connector	Black	2	Connectors → Connector blocks
#4 angle connector	Black	2	Connectors → Connector blocks

(continued)

table A-4: connectors (continued)

piece	color	quantity	classification
#6 angle connector	Black	16	Connectors → Connector blocks
Catch	Black	4	Connectors → Connector blocks
Extended catch	Medium stone gray	1	Connectors → Connector blocks
Axle extender	Dark stone gray	2	Connectors → Connector blocks
Peg extender	Black	4	Connectors → Connector blocks
Flexible axle joiner	Black	8	Connectors → Connector blocks
Double peg joiner	Black	4	Connectors → Connector blocks
Cornered peg joiner	Medium stone gray	1	Connectors → Connector blocks

table A-4: connectors (continued)

piece	color	quantity	classification
3M pegged block	Medium stone gray	13	Connectors → Connector blocks
5M pegged perpendicular block	Medium stone gray	8	Connectors → Connector blocks

table A-5: gears

piece	color	quantity	classification
8t gear	Medium stone gray	6	Gears → Spur gears
16t gear	Medium stone gray	2	Gears → Spur gears
24t gear	Medium stone gray	7	Gears → Spur gears
40t gear	Medium stone gray	1	Gears → Spur gears

(continued)

table A-5: gears (continued)

piece	color	quantity	classification
12t double bevel gear	Black	4	Gears → Double bevel gears
20t double bevel gear	Medium stone gray	2	Gears → Double bevel gears
36t double bevel gear	Black	1	Gears → Double bevel gears
Worm gear	Black	2	Gears → Other gears
Knob wheel	Black	4	Gears → Other gears
Turntable	Dark stone gray and black	1	Gears → Other gears

table A-6: miscellaneous elements

piece	color	quantity	classification
1 × 1 cone	White	3	Miscellaneous elements
TECHNIC tooth	Orange	8	Miscellaneous elements
Medium pulley wheel	Medium stone gray	2	Miscellaneous elements
TECHNIC pincer	Pearl gray	4	Miscellaneous elements
Balloon wheel	Medium stone gray	4	Miscellaneous elements

(continued)

table A-6: miscellaneous elements (continued)

piece	color	quantity	classification
Balloon tire	Black	4	Miscellaneous elements
Plastic ball	Red	1	Miscellaneous elements
Plastic ball	Blue	1	Miscellaneous elements

NXT-G quick reference

There are seven categories of NXT-G programming blocks and thirty types of standard programming blocks. Although Part III discusses how to use these blocks, this appendix gathers the basic facts about each one for easy reference. Table B-1 lists and describes the seven block categories. Tables B-2 through B-7 correspond to the first six categories and list the names, data hub characteristics, and functions of their respective blocks along with images of the blocks in their default configurations. I do not present a table for the seventh category, Custom blocks, because there are no standard blocks in this category.

NOTE Remember that you can access a block's data hub chart on the NXT software's documentation. See "Data Plug Characteristics" on page 92 for more information about this chart.

table B-1: NXT-G programming blocks

category	description
Common blocks	Includes the most frequently used programming blocks
Action blocks	Includes blocks that instruct your robot to carry out particular actions
Sensor blocks	Includes blocks that read a sensor and then send the value or related data through a data wire to another block
Flow blocks	Includes blocks that control the flow of a program
Data blocks	Includes blocks that process data and then send the output through a data wire to another block
Advanced blocks	Includes blocks that perform miscellaneous advanced functions
Custom blocks	Includes blocks created by the user (My Blocks) and those downloaded from the Internet (Web Blocks)

table B-2: common blocks

block	data hub	function
Move	Optional. Consists of number and logic plugs.	Controls one to three servo motors and specializes in driving simple mobile robots.
Record/Play	Optional. Consists of number, logic, and text plugs.	Records and plays movements of one to three servo motors.
Sound	Optional. Consists of number and text plugs.	Plays sound files and tones on the loudspeaker.
Display	Optional. Consists of number, logic, and text plugs.	Displays text, drawings, and images on the LCD.
Wait	No data hub.	Waits until a specified condition has been met before allowing the next block on the sequence beam to execute.
Loop	No data hub. Depending on the block's configuration, number and logic plugs may appear.	Repeats the code on its sequence beam until a specified condition has been met.

table B-2: common blocks (continued)

block	data hub	function
Switch	No data hub. Depending on the block's configuration, number, logic, or text plugs may appear.	Chooses between two or more paths of code (i.e., sequence beam paths) based on a specified condition.

table B-3: action blocks

block	data hub	function
Motor	Optional. Consists of number and logic plugs.	Specializes in controlling one servo motor.
Sound	See *Sound block* in Table B-2.	See *Sound block* in Table B-2.
Display	See *Display block* in Table B-2.	See *Display block* in Table B-2.
Send Message	Optional. Consists of number, logic, and text plugs.	Sends a wireless message formatted as text, number, or logic data to another NXT via Bluetooth.

table B-4: sensor blocks

block	data hub	function
Touch Sensor	Required. Consists of number and logic plugs.	Reads a touch sensor and tests for a specified condition: pressed, released, or bumped.
Sound Sensor	Required. Consists of number and logic plugs.	Reads a sound sensor and tests for a value that is greater than or less than a specified trigger point.
Light Sensor	Required. Consists of number and logic plugs.	Reads a light sensor and tests for a value that is greater than or less than a specified trigger point.
Ultrasonic Sensor	Required. Consists of number and logic plugs.	Reads an ultrasonic sensor and tests for a value that is greater than or less than a specified trigger point.
NXT Buttons	Required. Consists of number and logic plugs.	Reads the Enter, Left, or Right button and tests for a specified condition: pressed, released, or bumped.
Rotation Sensor	Required unless resetting the rotation sensor. Consists of number and logic plugs.	Reads the built-in rotation sensor in a servo motor and tests for a value that is greater than or less than a specified trigger point.
Timer	Required unless resetting the timer. Consists of number and logic plugs.	Reads one of three built-in timers and tests for a value that is greater than or less than a specified trigger point.
Receive Message	Required. Consists of number, logic, and text plugs.	Reads one of ten mailboxes for a message formatted as text, number, or logic data and can also compare the incoming message to a test message.

table B-5: flow blocks

block	data hub	function
Wait	See *Wait block* in Table B-2.	See *Wait block* in Table B-2.
Loop	See *Loop block* in Table B-2.	See *Loop block* in Table B-2.
Switch	See *Switch block* in Table B-2.	See *Switch block* in Table B-2.
Stop	Optional. Consists of a logic plug.	Stops the program as well as motors, sounds, and lamps.

table B-6: data blocks

block	data hub	function
Logic	Required. Consists of logic plugs.	Performs one of the following four logical operations on logic data: And, Or, Xor, or Not.
Math	Required. Consists of number plugs.	Performs one of the following four operations on number data: addition, subtraction, multiplication, or division.
Compare	Required. Consists of number and logic plugs.	Performs one of the following three comparisons on number data: less than, greater than, or equals.
Range	Required. Consists of number and logic plugs.	Performs a comparison on number data to see if it's inside or outside a range of specified numbers.
Random	Required. Consists of number plugs.	Generates a random number between specified minimum and maximum limits, inclusive.
Variable	Required. Consists of a number, logic, or text plug, depending on the block's configuration.	Writes number, logic, or text data to variables and also retrieves data from existing variables.

table B-7: advanced blocks

block	data hub	function
Text	Required. Consists of text plugs.	Combines separate pieces of text data into one piece of text data.
Number to Text	Required. Consists of number and text plugs.	Converts number data into text data so that it can be displayed on the LCD.
Keep Alive	Optional. Consists of a number plug.	Prevents the NXT from turning off by overriding its Sleep mode settings.
File Access	Optional or required, depending on block's configuration. Consists of number, logic, and text plugs.	Stores and reads text and number data files. These files are retained even when you turn off the NXT.
Calibrate	Optional. Consists of number and logic plugs.	Calibrates a light or sound sensor by setting its minimum and maximum value.
Reset Motor	Optional. Consists of logic plugs.	Resets an automatic error-correction mechanism in the servo motors.

internet resources

The magnitude of Internet resources for MINDSTORMS and LEGO in general can provide a great source of inspiration and help you hone your building and programming skills in a variety of ways. In this appendix, I've gathered a number of popular LEGO websites (both official and unofficial), but you can easily discover additional ones by following any links to related sites or doing a Google search.

NOTE You can find up-to-date links at this book's website: http://www.nxtguide.davidjperdue.com.

general resources

LEGO MINDSTORMS (http://www.mindstorms.com)
The official MINDSTORMS website should be your first stop. Here you'll find a variety of resources including an FAQ, additional building instructions, important software updates (http://mindstorms.lego.com/support/updates), advanced resources for developers (http://mindstorms.lego.com/overview/nxtreme.aspx), and much more.

NXTLOG (http://www.mindstorms.com/nxtlog)
A free feature of the official MINDSTORMS website, NXTLOG offers users of the NXT set a place to display their robots, browse and comment on thousands of other creations, and even compete in contests!

LUGNET (http://www.lugnet.com)
The LEGO Users Group Network (LUGNET) is an unofficial website that offers a number of different forums (including an active robotics forum), an interactive map for finding a LEGO users group in your area, and many other valuable resources.

MOCpages (http://www.mocpages.com)
MOCpages (short for *My Own Creation Pages*) is a site where you can display your LEGO creations for free (not just NXT creations). You also can create and customize your own home page, write reviews of other people's creations, and more. Make sure to check out the TECHNIC category and robotics subcategory of creations.

Brickshelf (http://www.brickshelf.com)
Brickshelf offers free image hosting for your LEGO creations. If you need some inspiration for your latest creation, try browsing the thousands of incredible pictures!

Peeron LEGO Inventories (http://www.peeron.com)
Peeron features a detailed database of LEGO sets and parts—you'll even find scans of official building instructions.

Brickset (http://www.brickset.com)
Brickset is a comprehensive guide to LEGO sets with links to websites where you can purchase the sets.

MINDSTORMS NXT Building Instructions (http://www.freewebs.com/legorobots)
This site offers free building instructions for several NXT models made by LEGO fans. A brick sorter and robotic dog are just two of the creations here.

Technica (http://isodomos.com/technica/technica.html)
Jim Hughes has put together a fantastic site that not only provides the history of the LEGO TECHNIC series but also lists most TECHNIC sets and pieces with pictures and detailed information.

programming resources

Here are web addresses for some important and useful programming resources for the NXT. If you're looking for a program that can convert sound files on your computer to the format used by the NXT, visit the Programmable Brick Utilities web page and look for a utility called Wav2Rso.

NBC and NXC
http://bricxcc.sourceforge.net/nbc

NBC Debugger for NXT
http://www.sorosy.com/lego/nxtdbg

BricxCC
http://bricxcc.sourceforge.net

Programmable Brick Utilities
http://bricxcc.sourceforge.net/utilities.html

leJOS NXJ
http://lejos.sourceforge.net

RobotC
http://www.robotc.net

Writing Efficient NXT-G Programs
http://www.firstlegoleague.org/sitemod/upload/Root/WritingEfficientNXTGPrograms2.pdf

OnBrick NXT Remote Control for PDAs and PCs
http://www.pspwp.pwp.blueyonder.co.uk/science/robotics/nxt

NXTender
http://www.tau.ac.il/~stoledo/lego/NXTender

NXT Programming Software
http://www.teamhassenplug.org/NXT/NXTSoftware.html

bluetooth resources

If you want to find out more about how to use Bluetooth with your NXT set, these online resources will get you started. In addition, you can find out more about the recommended D-Link DBT-120 Bluetooth adapter at the D-Link website given below.

MINDSTORMS Bluetooth Resources
http://www.mindstorms.com/bluetooth

NXT Bluetooth Compatibility List
http://www.vialist.com/users/jgarbers/NXTBluetoothCompatibilityList

Analysis of the NXT Bluetooth-Communication Protocol
http://www.tau.ac.il/~stoledo/lego/btperformance.html

D-Link
http://www.dlink.com

NXT blogs

NXT blogs have become an important part of the MINDSTORMS community by providing news, projects, reviews, and other great resources for users of the NXT set. These are some of the most popular NXT blogs that have appeared since the NXT set's release.

The NXT STEP
http://www.thenxtstep.com

nxtasy.org
http://www.nxtasy.org

bNXT
http://www.bnxt.com

LEGO computer-aided design resources

There are now a variety of computer-aided design (CAD) programs (most of which are free) with which you can create virtual NXT models, although the LDraw system of tools—which I used for this book—is probably the most powerful. All of the websites listed after LDraw.org are either specifically for LDraw or related to LDraw in some way.

LEGO Digital Designer
http://ldd.lego.com

Google SketchUp NXT Parts Library
http://groups.google.com/group/
LegoTechnicandMindstormsNXTParts

Solid Modeling
http://www-education.rec.ri.cmu.edu/solidmodel

LDraw
http://www.ldraw.org

MLCAD
http://www.lm-software.com/mlcad

LeoCAD
http://www.leocad.org

Bricksmith
http://bricksmith.sourceforge.net

LPub
http://www.kclague.net/LPub

LSynth
http://www.kclague.net/LSynth

L3P
http://www.hassings.dk/l3/l3p.html

LDView
http://ldview.sourceforge.net

building techniques

For more information on building techniques (including how to build with TECHNIC bricks), visit these web pages.

TECHNIC Design School
http://technic.lego.com/technicdesignschool

LEGO Design School
http://creator.lego.com/designschool

LEGO Education Constructopedia
http://www.lego.com/education/default.asp?l2id=3_
3&page=4_1

Understanding LEGO Geometry
http://www.syngress.com/book_catalog/174_lego_robo/
chapter_01.htm

LEGO Design
http://www.owlnet.rice.edu/~elec201/Book/legos

Sergei Egorov's LEGO Geartrains
http://www.malgil.com/esl/lego/geartrains.html

educational resources

If you're interested in using LEGO MINDSTORMS NXT as an educational tool, the following websites will be of interest to you.

LEGO Education
http://www.legoeducation.com

LEGO ED West
http://www.legoedwest.com

LEGO Engineering
http://www.legoengineering.com

FIRST LEGO League
http://www.firstlegoleague.org

US FIRST Curriculum Collection
http://www.usfirst.org/community/fll/content.aspx?id=798

Robotics Academy
http://www-education.rec.ri.cmu.edu

LEGO sets, LEGO pieces, and custom hardware

These web addresses are for both official and unofficial sites that sell LEGO sets and pieces, or custom hardware made for the NXT set (such as custom sensors).

LEGO Store
http://shop.lego.com

LEGO Education Store
http://www.legoeducation.com/store

BrickLink
http://www.bricklink.com

HiTechnic
http://www.hitechnic.com

Mindsensors.com
http://www.mindsensors.com

LEGO storage

These sites offer tackle boxes and other organizers that you can use to store your LEGO pieces.

Robotics Learning Store
http://www.roboticslearning.com/store

Storage and Organizers at the LEGO Education Store
http://www.legoeducation.com/store/SearchResult.aspx?pt=17

Plano Molding Company
http://www.planomolding.com

personal websites

A number of MINDSTORMS fans have personal websites (or web pages) that feature their creations and other related LEGO material. While not all of the following sites include NXT material specifically, all of them do have useful MIND-STORMS information.

David J. Perdue
http://www.davidjperdue.com

Philippe Hurbain
http://www.philohome.com

Dave Astolfo
http://www.astolfo.com

Daniele Benedettelli
http://daniele.benedettelli.com

Michael Gasperi
http://extremenxt.com/lego.htm

Matthias Paul Scholz
http://mynxt.matthiaspaulscholz.eu

Steve Hassenplug
http://www.teamhassenplug.org

Brian Davis
http://mindstorms.lego.com/MeetMDP/BDavis.aspx

Bryan Bonahoom
http://www.funtimetechnologies.com/teamb2

Laurens Valk
http://www.freewebs.com/laurens200

Jürgen Stuber
http://www.jstuber.net

Kevin Clague
http://www.kclague.net

Mario Ferrari
http://www.marioferrari.org/lego.html

Miguel Agullo
http://technicpuppy.miguelagullo.net

LEGO events

These websites are for several large LEGO events that usually occur annually. Sometimes MINDSTORMS contests are held during these events.

LEGO World
http://www.legoworld.nl

BrickFest
http://www.brickfest.com

BrickWorld
https://registration.brickworld.us

NWBrickCon
http://www.nwbrickcon.org

index

The Unofficial LEGO MINDSTORMS NXT Inventor's Guide is set in Chevin.

The book was printed and bound at Malloy Incorporated in Ann Arbor, Michigan. The paper is 60# Spring Forge Smooth, which is certified by the Sustainable Forestry Initiative (SFI). The book uses a RepKover binding, which allows it to lay flat when open.

The LEGO® MINDSTORMS® NXT Idea Book
Design, Invent, and Build

by MARTIJN BOOGAARTS, JONATHAN A. DAUDELIN, BRIAN L. DAVIS, JIM KELLY, DAVID LEVY, LOU MORRIS, FAY RHODES, RICK RHODES, MATTHIAS PAUL SCHOLZ, CHRISTOPHER R. SMITH, *and* ROB TOROK

With chapters on programming and design, CAD-style drawings, and abundance of screenshots, *The LEGO MINDSTORMS NXT Idea Book* makes it easy for readers to master the LEGO MIND-STORMS NXT kit and build the nine example robots. Readers learn about the NXT parts (beams, axles, gears, and so on) and how to combine them to build and program working robots like a slot machine (complete with flashing lights and a lever), a black-and-white scanner, and even a robot DJ. Chapters cover using the NXT programming language (NXT-G) as well as trouble-shooting, software, sensors, Bluetooth, and even how to create a NXT remote control. LEGO fans of all ages will find this book an ideal jumping-off point for doing more with the NXT kit.

SEPTEMBER 2007, 368 PP., $29.95 ($35.95 CDN)
ISBN 978-1-59327-150-3

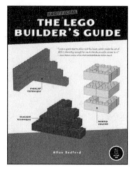

The Unofficial LEGO® Builder's Guide
by ALLAN BEDFORD

The Unofficial LEGO Builder's Guide combines techniques, principles, and reference information for building with LEGO bricks that go far beyond LEGO's official product instructions. Readers discover how to build everything from sturdy walls to a basic sphere, as well as projects including a mini space shuttle and a train station. The book also delves into advanced concepts such as scale and design. Includes essential terminology and the Brickopedia, a comprehensive guide to the different types of LEGO pieces.

SEPTEMBER 2005, 344 PP., $24.95 ($33.95 CDN)
ISBN 978-1-59327-054-4

Getting Started with LEGO® Trains
by JACOB H. MCKEE

Getting Started with LEGO Trains shows you everything from how to set up train tracks to how to build custom freight cars. LEGO insider Jake McKee shares some of his most fascinating and original train designs, while including descriptive articles on basic building techniques and high-quality building instructions for several different projects. For veteran LEGO trains fans and curious beginners.

MARCH 2004, 120 PP., $14.95 ($21.95 CDN)
ISBN 978-1-59327-006-3